DEMIS HASSABIS AND THE NOBEL PRIZE HANDBOOK

Creating a Legacy of Interdisciplinary Scientific Breakthroughs

SEAN T. ROLAND

COPYRIGHT

Copyright©2024 Sean T. Roland. All rights reserved. No part of this publication may be reproduced, distributed, or transmitted in any form or by any means, including photocopying, recording, or other electronic or mechanical methods, without the prior written permission of the publisher, except in the case of brief quotations embodied in critical reviews and certain other non-commercial uses permitted by copyright law

TABLE OF CONTENTS

COPYRIGHT 1
TABLE OF CONTENTS 2
INTRODUCTION 4
    The Life, Work, and Nobel Prize Legacy of Demis Hassabis 4
CHAPTER 1 11
    Early Life and Academic Foundations of Demis Hassabis 11
CHAPTER 2 19
    The Formation of DeepMind: Vision and Beginnings .. 19
CHAPTER 3 29
    AlphaGo: The World's First Superhuman AI in Strategy Games 29
CHAPTER 4 39
    AlphaFold: Revolutionizing Protein Structure Prediction 39
CHAPTER 5 53
    The 2024 Nobel Prize: Recognition of Breakthroughs in Science 53
CHAPTER 6 64
    The Broader Implications of AI in Science and Medicine 64
CHAPTER 7 77
    AI Ethics and the Future of Machine Learning 77

CHAPTER 8 88

   Collaborations with David Baker and John Jumper 88

CHAPTER 9 99

   The Role of Artificial Intelligence in Future Scientific Discoveries 99

CHAPTER 10 110

   Demis Hassabis' Legacy and Future Prospects 110

CONCLUSION 121

   Demis Hassabis and the Nobel Prize – A Legacy in the Making 121

INTRODUCTION

The Life, Work, and Nobel Prize Legacy of Demis Hassabis

Demis Hassabis stands as one of the most visionary figures of the 21st century, with his contributions to **artificial intelligence (AI)** and **biological sciences** culminating in the **2024 Nobel Prize in Chemistry**. Co-founder and CEO of **DeepMind**, Hassabis has spearheaded groundbreaking research that has bridged the gap between AI and complex scientific problems, fundamentally reshaping our understanding of both artificial intelligence and biological systems.

Hassabis' Nobel Prize was shared with **John Jumper** and **David Baker** for their groundbreaking work in **protein structure prediction**, a field of biology that has long challenged scientists. The team's development of **AlphaFold**, an AI-powered system that accurately predicts the 3D structure of proteins from their amino acid sequences, represents a historic achievement. This monumental breakthrough addresses the so-called "**protein folding problem**," which has perplexed scientists for over half a century. The ability to understand how proteins fold is essential in numerous fields, from **medicine** to **biotechnology**, and their success in this area has set the stage for advancements in **drug discovery**, **vaccine development**, and **biomedical research**.

This introduction will outline Hassabis' remarkable journey from his early years as a chess prodigy and video game developer to becoming one of the most influential AI researchers of our time. We will delve into his formative years, his academic and professional achievements, and the key moments that led to his winning the **Nobel Prize in Chemistry**. Moreover, the introduction will set the stage for an in-depth exploration of how Hassabis' work has revolutionized **scientific research** and **biomedical applications**.

Early Life and Intellectual Foundations

Born in **London** in 1976, Demis Hassabis displayed extraordinary intellectual promise from a young age. His early years were marked by a passion for chess, reaching the level of **chess master** by the age of 13. This keen ability to think strategically would later influence his approach to artificial intelligence and problem-solving. While chess introduced him to the complexity of decision-making and pattern recognition, his early interest in **video games** became another important avenue for exploring the boundaries of AI.

Hassabis' education is reflective of his multidisciplinary curiosity. He earned a degree in **Computer Science** from **Cambridge University**, where he studied alongside some of the brightest minds in computing. After Cambridge, Hassabis pursued a PhD in **Cognitive Neuroscience** at **University College London (UCL)**, where his research focused on understanding memory and decision-making processes in the brain. This deep dive into how the human mind works would eventually inspire Hassabis to explore the creation of artificial systems that mimic human intelligence, combining insights from **neuroscience** with **machine learning**.

The Formation of DeepMind: A Vision for General AI

In 2010, Hassabis co-founded **DeepMind** with the mission of solving intelligence. His goal was to create **artificial general intelligence (AGI)**—a system that could learn and perform any intellectual task that a human could, rather than being limited to specific domains. The early work of DeepMind focused on applying **deep learning** and **reinforcement learning** to various problems, including game-playing AI. One of DeepMind's early successes was

AlphaGo, an AI system that defeated world champion Lee Sedol in the ancient game of **Go**—a feat that many considered impossible for machines, given the game's complexity.

AlphaGo's victory in 2016 signaled a major leap forward in **AI development**, demonstrating that machine learning systems could surpass human experts in tasks that require intuition, pattern recognition, and long-term strategy. This success brought Hassabis and DeepMind into the spotlight, drawing attention to the potential of **AI systems** to address real-world problems beyond games, such as health care, energy, and biology.

AlphaFold: Solving the Protein Folding Problem

While AlphaGo was a groundbreaking achievement in the world of AI, Hassabis and his team at DeepMind soon turned their attention to a more pressing scientific challenge: the **protein folding problem**. Proteins are the molecular machines that perform nearly all biological functions, and understanding their structure is key to unlocking their function. However, predicting the three-dimensional structure of a protein from its amino acid sequence had been an unsolved problem in **biology** for more than 50 years.

In 2020, DeepMind unveiled **AlphaFold**, an AI system that could predict protein structures with astonishing accuracy. The system used a combination of **neural networks** and **deep learning** algorithms, trained on vast datasets of known protein structures, to model the folding patterns of proteins. The breakthrough was recognized as one of the most significant scientific achievements of the decade, earning DeepMind and Hassabis widespread acclaim.

AlphaFold's success in the **CASP14 competition** (Critical Assessment of Protein Structure Prediction) was a turning point in **biomedical research**. For the first time, scientists had a tool that could predict protein structures in days rather than years, dramatically accelerating research in **drug discovery**, **vaccine development**, and understanding diseases like **Alzheimer's**, **cancer**, and **Parkinson's**.

The Nobel Prize: Recognizing the Intersection of AI and Biology

In 2024, the **Nobel Prize in Chemistry** was awarded to **Demis Hassabis**, **John Jumper**, and **David Baker** for their work on **protein structure prediction**. This Nobel Prize was not just a recognition of AlphaFold's technical prowess; it was a landmark moment in the intersection of **AI** and **biology**. For the first time, a Nobel Prize honored the role of

AI-driven research in solving one of biology's greatest mysteries.

The Nobel Prize highlighted the potential of AI to contribute meaningfully to scientific discovery. What once seemed like science fiction—the ability for machines to understand and predict the behavior of biological systems—was now a reality. This achievement has profound implications for **medicine**, **biotechnology**, and **environmental science**.

AlphaFold's ability to predict protein structures has led to breakthroughs in **drug design**, allowing researchers to develop treatments more quickly and with greater precision. Its impact on vaccine development, especially during the **COVID-19 pandemic**, has further demonstrated the power of AI in addressing global health crises. By predicting the structure of viral proteins, AlphaFold has helped scientists design more effective vaccines and therapies.

The Future of AI in Science and Medicine

Demis Hassabis' journey does not end with the Nobel Prize. His work with **DeepMind** and his vision for the future of **AI** continue to drive the development of technologies that will transform how we understand and interact with the world. The success of **AlphaFold** is just the beginning of what AI can achieve in **biology**. Looking forward, Hassabis envisions a future where **AI systems** assist in solving the greatest scientific challenges, from **climate change** to **neuroscience**.

Hassabis has been vocal about the ethical implications of AI, advocating for responsible and transparent development of AI technologies. He recognizes that while AI holds immense potential, it must be used ethically to avoid the risks of misuse or unintended consequences. As AI becomes more integrated into scientific research, the need for clear ethical guidelines will only become more important.

Demis Hassabis' Nobel Prize in Chemistry represents the culmination of decades of innovation, collaboration, and intellectual curiosity. From his early work in **video games** and **chess** to his leadership at **DeepMind** and the development of **AlphaFold**, Hassabis has consistently pushed the boundaries of what AI can achieve. His work has not only advanced the field of **artificial intelligence** but has also provided powerful tools for solving some of the most challenging problems in **biomedicine** and **biotechnology**.

As we continue to explore the implications of Hassabis' work, it becomes clear that **AI** is not just a tool for improving computational efficiency—it is a transformative force that is reshaping the landscape of **scientific discovery**.

CHAPTER 1
Early Life and Academic Foundations of Demis Hassabis

Demis Hassabis, a pioneering figure in **artificial intelligence (AI)**, is known for his significant contributions to AI research and its applications in solving some of the world's most complex problems. His early life and academic journey played a crucial role in shaping the intellectual curiosity and interdisciplinary thinking that would later drive his groundbreaking work. This chapter explores Hassabis' **upbringing**, his educational background, and how early experiences with **chess** and **video games** profoundly influenced his approach to artificial intelligence.

Upbringing in London and Mixed Heritage

Demis Hassabis was born in **London** on **July 27, 1976**, to a family with a rich cultural heritage. His father, a **Cypriot-Greek**, and his mother, of **Chinese-Singaporean descent**, provided a multicultural environment that instilled in Hassabis a deep appreciation for different perspectives and problem-solving approaches. Growing up in **North London**, Hassabis was exposed to a range of intellectual and cultural experiences that would later inspire his career in science and technology.

His mixed heritage gave him a unique outlook on life, allowing him to draw from both **Eastern** and **Western** traditions. The multicultural environment he grew up in fostered **open-mindedness** and a **global perspective**, which became evident in his later work, where he focused on using AI to address global challenges like healthcare and climate change.

Hassabis was recognized early on for his exceptional intellect and curiosity. His natural talents, combined with a supportive family environment, allowed him to explore a wide range of interests. His father, an art lover, encouraged Hassabis to think creatively, while his mother, a librarian, provided an environment rich in knowledge and learning. From a young age, Hassabis exhibited a fascination with both the arts and sciences, a dual interest that would later define his interdisciplinary approach to AI research.

Chess Prodigy and Early Strategic Thinking

One of the most defining aspects of Hassabis' early life was his proficiency in **chess**. By the age of four, he was already showing signs of extraordinary intelligence, and by the age of 13, Hassabis had become a **chess master**. His prodigious talent for the game, known for its reliance on **strategy**, **pattern recognition**, and **long-term planning**, was not only a testament to his intellectual abilities but also foreshadowed his later work in artificial intelligence. Chess, after all, is often considered a prime testing ground for AI due to its complexity and the need for strategic thinking.

His experiences as a young chess player helped him develop a deep understanding of **decision-making processes** and **strategic planning**, skills that would later become central to his AI research. Chess taught him to think many steps ahead, anticipate his opponent's moves, and make decisions based on incomplete information—all critical components of AI systems designed to solve real-world problems.

Hassabis' early achievements in chess were recognized by the broader chess community, and he regularly competed against much older and more experienced players. His prowess in the game caught the attention of chess enthusiasts and researchers alike, who noted the parallels between chess and the emerging field of **machine learning**. Though his career as a chess player would not last long, the skills he honed during his time as a chess master would profoundly influence his later work in AI.

Passion for Video Games and Early Programming

Alongside chess, Hassabis had another childhood passion that would shape his intellectual journey: **video games**. As a child, he spent countless hours playing video games, fascinated not only by the gameplay but also by the mechanics behind them. By the age of eight, Hassabis began experimenting with **game programming**, and by his teenage years, he was designing games himself.

At the age of **17**, he joined **Bullfrog Productions**, a well-known video game studio, where he worked on the hit game **Theme Park**. His experience in video game design sharpened his ability to work with **complex systems** and **simulations**, skills that would later prove invaluable in his AI research. Video games offered a unique environment in which players could engage with intricate systems and solve problems, elements that strongly influenced how Hassabis would later approach **artificial intelligence**.

His time in the gaming industry also exposed him to the concept of **simulated worlds**—virtual environments governed by sets of rules, where complex behaviors emerge from simple algorithms. This exposure would later inspire Hassabis to develop AI systems that could simulate human cognition and decision-making. Video games, in many ways, were Hassabis' first laboratory for understanding how **intelligence** works and how it could be simulated in machines.

Academic Journey: Cambridge and UCL

Hassabis' academic journey was equally impressive. After excelling in secondary education, he gained admission to **Cambridge University**, where he studied **computer science** at **Queens' College**. Cambridge, known for its rigorous academic environment and intellectual community, provided the perfect setting for Hassabis to explore his growing interest in **AI** and **machine learning**. During his time at Cambridge, Hassabis not only excelled academically but also honed his problem-solving skills by working on complex computational challenges.

After graduating from Cambridge, Hassabis' intellectual curiosity led him to pursue a PhD in **Cognitive Neuroscience** at **University College London (UCL)**. At UCL, he worked under the guidance of Professor **Eleanor Maguire**, a leading expert in memory and brain function. His doctoral research focused on **memory** and **decision-making**, with particular emphasis on how the brain constructs **internal models** of the world to guide behavior.

Hassabis' time at UCL was transformative. By studying the brain's **neural mechanisms** for memory and learning, he gained a deep understanding of how biological systems process information—a key insight that would later inform his work in developing AI systems that mimic human intelligence. His research at UCL earned him critical acclaim, and he published several influential papers on **episodic memory** and the **hippocampus**, a brain region involved in memory formation.

It was during his time at UCL that Hassabis began to draw connections between the **biological brain** and the **artificial mind**. He realized that many of the principles governing human cognition, such as learning from experience and adapting to new environments, could be applied to AI systems. This insight would become a driving force behind his later work in **deep learning** and **reinforcement learning**—two key areas of AI research.

How Chess and Video Games Influenced His Approach to AI

Chess and video games were more than childhood passions for Hassabis; they became integral to his intellectual framework for understanding and developing AI. Both activities taught him valuable lessons about **strategy**,

pattern recognition, and **problem-solving**, skills that are essential for building intelligent systems.

In chess, Hassabis learned the importance of **search algorithms**—techniques used to explore possible moves and evaluate their outcomes. These same principles underpin many AI systems, particularly those involved in **decision-making** and **planning**. His experience in video game design, meanwhile, gave him a deep appreciation for **simulated environments**—virtual worlds governed by rules and algorithms that mirror real-world complexity. This understanding of simulations became crucial when Hassabis and his team at **DeepMind** developed **AlphaGo** and **AlphaFold**, AI systems that rely on deep learning to simulate complex processes.

Moreover, both chess and video games provided Hassabis with an early understanding of how **artificial systems** could be used to model and predict behavior. In chess, the challenge was to anticipate an opponent's moves, while in video games, it was about creating systems that responded to players' actions in real-time. These insights would later inspire Hassabis to develop AI systems that could learn from experience and adapt to new information, just as humans do.

Demis Hassabis' early life and academic foundations set the stage for his groundbreaking work in **artificial intelligence**. His experiences as a **chess master**, **video game designer**, and **neuroscientist** provided him with a unique set of skills and insights that would later inform his approach to AI research. Through his academic journey at **Cambridge** and **UCL**, Hassabis developed a deep understanding of how **intelligence**—both biological and artificial—can be understood, simulated, and applied to solve real-world problems.

As we continue to explore Hassabis' contributions to AI, it is clear that his early influences in chess, video games, and neuroscience played a crucial role in shaping his vision for **DeepMind** and its mission to solve intelligence. His interdisciplinary approach, combining **neuroscience**, **computer science**, and **AI**, has not only revolutionized the field of artificial intelligence but also earned him the **2024 Nobel Prize in Chemistry** for his role in solving the **protein folding problem** through **AlphaFold**. This early foundation will remain a central theme as we delve deeper into his scientific career and its far-reaching impact on the future of technology and biology.

CHAPTER 2

The Formation of DeepMind: Vision and Beginnings

The story of **DeepMind Technologies** is one of the most significant milestones in the field of **artificial intelligence (AI)**. Founded in 2010 by **Demis Hassabis, Shane Legg**, and **Mustafa Suleyman**, DeepMind began with an ambitious vision: to combine **neuroscience** and **machine learning** to develop AI systems that mimic the learning processes of the human brain. The journey from an early-stage AI research company to becoming a subsidiary of **Google** in 2014 has cemented DeepMind's place as a global leader in artificial intelligence, with landmark achievements such as **AlphaGo** and **AlphaFold**. In this chapter, we will explore the formation of DeepMind, its early vision, key milestones, and its eventual acquisition by Google, which allowed it to expand its research and apply AI to even broader challenges.

The Birth of DeepMind: Co-Founders and Initial Vision

In 2010, **Demis Hassabis**, along with **Shane Legg** and **Mustafa Suleyman**, co-founded **DeepMind Technologies** in London. Hassabis brought his extensive background in **neuroscience** and **computer science** to the project, while Shane Legg, a machine learning expert, and Mustafa Suleyman, an entrepreneur focused on the social implications of AI, added complementary skills to the team. Their shared vision was to create an artificial intelligence capable of **general-purpose learning**—one that could adapt, learn from experience, and eventually tackle a wide variety of tasks, much like the human brain.

Hassabis had long been interested in **human cognition** and how the brain learns and makes decisions. His doctoral work in **cognitive neuroscience** at **University College London (UCL)** provided him with insights into the neural mechanisms behind memory, learning, and planning, which he believed could be replicated in machines. Meanwhile, Shane Legg had worked extensively in **machine learning**, focusing on algorithms that could enable machines to learn from data. Together, their expertise laid the foundation for an approach that blended **neuroscience** with cutting-edge **machine learning algorithms**.

The original vision of DeepMind was to build **artificial general intelligence (AGI)**, an AI system that could learn and solve any intellectual task a human could, regardless of domain. AGI differed from traditional AI approaches, which typically focused on creating algorithms for specific tasks, such as image recognition or language translation. Hassabis and his team believed that by drawing inspiration from how humans learn, they could create AI that could apply **generalizable knowledge** across multiple domains, making it much more versatile and powerful than task-specific AI systems.

DeepMind's mission statement—"to solve intelligence and then use that to solve everything else"—reflected their belief that creating human-level AI could unlock solutions to some of the most challenging problems in areas like **healthcare**, **climate change**, and **energy**

Early Years: Neuroscience Meets Machine Learning

In its early years, DeepMind's research focused heavily on applying insights from **neuroscience** to **reinforcement learning**, a type of machine learning that allows AI systems to learn from trial and error, much like humans. Reinforcement learning involves an agent that interacts with an environment, receiving **rewards** or **penalties** for actions taken, and adjusting its behavior accordingly to maximize its cumulative reward. This learning method was inspired by how the brain's reward system reinforces behaviors that lead to positive outcomes.

DeepMind's innovation lay in combining **deep neural networks** with reinforcement learning to create **deep reinforcement learning**. Neural networks, modeled loosely on the structure of the human brain, allowed the AI to process large amounts of data, recognize patterns, and make decisions based on those patterns. By integrating this capability with reinforcement learning, DeepMind was able to create systems that could learn from experience in complex environments—an approach that had never been fully realized before.

A major goal of the company was to develop AI systems that could **learn autonomously** and **generalize** their knowledge across different tasks. This meant that instead of building a new AI for each specific problem, the same AI could be trained on one task and apply its learning to other related tasks, demonstrating true **general intelligence**.

First Milestones: Game-Playing AI and Breakthrough Successes

One of DeepMind's first major breakthroughs came with the development of **Atari-playing AI agents**. In 2013, the DeepMind team trained an AI using deep reinforcement learning to play several classic Atari 2600 video games, including **Pong**, **Breakout**, and **Space Invaders**. The AI started with no prior knowledge of the games and learned purely through trial and error by interacting with the game environment and adjusting its behavior based on feedback from the game score.

Remarkably, the AI was able to master many of these games, achieving superhuman performance in several of them. The key innovation was that the same AI could be used across multiple games without any modifications to the underlying algorithm—this was a major step toward AGI. The results were published in a highly influential paper titled "Playing Atari with Deep Reinforcement Learning" and garnered widespread attention from both the AI research community and the media.

The success of the Atari project demonstrated the power of DeepMind's approach to **deep learning** and **reinforcement learning**, and it attracted the attention of **Google**. At the time, Google was increasingly interested in AI, seeing its potential to revolutionize areas such as **search**, **advertising**, and **robotics**. Impressed by the achievements of DeepMind, Google initiated discussions about acquiring the company to further its own AI ambitions.

Acquisition by Google in 2014: A Strategic Move

In **2014**, Google acquired DeepMind for a reported **$500 million**. The acquisition marked a pivotal moment for both companies. For Google, acquiring DeepMind gave the tech giant access to some of the most talented AI researchers in the world and allowed it to deepen its focus on **AI research**. For DeepMind, the acquisition provided the resources and infrastructure necessary to scale its research efforts and tackle even more ambitious challenges.

As part of the acquisition, DeepMind negotiated terms with Google to ensure that their AI research would be used ethically. One of the conditions was the establishment of an **AI ethics board** to oversee the use of AI technologies developed by DeepMind, reflecting the team's concerns about the potential misuse of AGI. This focus on **ethical AI development** has remained a core part of DeepMind's mission, especially as AI continues to grow in power and influence across society.

The acquisition by Google allowed DeepMind to expand its research team, invest in **computational infrastructure**, and collaborate with other parts of Google to explore how AI could improve services like **Google Search** and **Google Assistant**. However, DeepMind maintained a significant degree of autonomy within Google, continuing to operate as an independent research lab with a mission focused on solving intelligence.

Post-Google Acquisition: AlphaGo and AI Breakthroughs

After the acquisition, DeepMind achieved some of its most well-known milestones, including the development of **AlphaGo**, the AI system that defeated world champion **Lee Sedol** in the ancient game of **Go** in 2016. Go had long been considered one of the most challenging games for AI to master due to its complexity and the vast number of possible board configurations. AlphaGo's victory over Sedol marked a historic moment in AI research, demonstrating that machines could now surpass human experts in tasks requiring intuition and strategic thinking.

AlphaGo was built using a combination of **deep neural networks** and **Monte Carlo Tree Search**, a search algorithm that helped the AI evaluate possible moves and outcomes. The system was trained using **supervised learning** on data from human games, as well as through **reinforcement learning**, where AlphaGo played millions of games against itself to improve its performance. AlphaGo's success captured the world's attention and cemented DeepMind's reputation as a leader in AI research.

The development of AlphaGo and its successors, **AlphaGo Zero** and **AlphaZero**, demonstrated the potential of AI to tackle complex problems in ways that went beyond traditional rule-based approaches. Unlike earlier versions of Go-playing AI, which relied on handcrafted heuristics and domain-specific knowledge, AlphaGo's strength lay in its ability to **learn autonomously** from data, using general-purpose algorithms that could be applied to other domains.

Looking Ahead: DeepMind's Broader Impact and Ambitions

With Google's support, DeepMind has continued to push the boundaries of AI research, applying its technology to areas beyond games. One of its most significant achievements came in 2020 with the development of **AlphaFold**, an AI system capable of predicting the 3D structure of proteins from their amino acid sequences. This breakthrough solved a long-standing challenge in **biology**, known as the **protein folding problem**, and has the potential to revolutionize fields like **drug discovery** and **biomedical research**.

DeepMind's broader mission remains focused on **solving intelligence** and using AI to tackle some of the world's most pressing challenges, from **healthcare** to **climate change**. Its continued work in areas like **reinforcement learning**, **AI ethics**, and **general-purpose AI** is likely to have far-reaching implications for both science and society.

The formation of **DeepMind** in 2010, led by **Demis Hassabis, Shane Legg,** and **Mustafa Suleyman,** marked the beginning of a bold new chapter in the history of artificial intelligence. From its early vision of combining **neuroscience** and **machine learning** to the groundbreaking successes of **Atari-playing AI** and **AlphaGo**, DeepMind has consistently demonstrated the power of AI to transform science and technology. The company's ability to combine insights from **neuroscience** with cutting-edge **machine learning** algorithms has enabled it to create **general-purpose AI** systems that have surpassed human experts in complex tasks like playing **Go** and solving the **protein folding problem**.

The acquisition by Google in 2014 allowed DeepMind to expand its resources and accelerate its research efforts, resulting in landmark achievements like **AlphaGo** and **AlphaFold**. DeepMind's mission to use AI to solve global challenges remains at the forefront of its work, and the company continues to push the boundaries of what AI can achieve in fields ranging from healthcare to environmental sustainability.

CHAPTER 3

AlphaGo: The World's First Superhuman AI in Strategy Games

In March 2016, **AlphaGo**, an artificial intelligence developed by **DeepMind**, made headlines worldwide by defeating **Lee Sedol**, a world champion in the ancient game of **Go**. Go, a board game that originated in China more than 2,500 years ago, had long been considered one of the most challenging games for computers to master due to its complexity and vast number of possible moves. AlphaGo's victory over Lee Sedol was not just a milestone in the history of AI; it was a major leap in **AI capabilities**, demonstrating the power of **deep neural networks** and **reinforcement learning**. This chapter chronicles the development of AlphaGo, analyzes the technological advancements it represented, and discusses the cultural and technological impact of this breakthrough on both the AI and gaming worlds.

The Significance of Go and Its Challenge for AI

To understand the significance of AlphaGo's victory, it's important to appreciate the complexity of the game itself. Go is often regarded as the ultimate strategy game due to its simplicity in rules but complexity in gameplay. The objective is to place stones on a board in such a way as to control more territory than your opponent. While the basic rules are easy to learn, the strategic depth of the game is vast.

What makes Go particularly challenging for computers is the sheer number of possible board configurations. For example, while chess has an estimated 10^{120} possible positions, Go has around 10^{170}—a number far exceeding the total number of atoms in the observable universe. Traditional **rule-based systems** that had been successful in chess-playing AI, such as **IBM's Deep Blue**, were not adequate for Go, as it was impossible to calculate every potential move due to the game's branching complexity.

In the past, attempts to develop Go-playing AI had relied on **brute-force search algorithms**, which were effective in simpler games like chess. However, due to Go's complexity, these approaches were insufficient. To master Go, an AI would need to incorporate **intuition** and **pattern recognition**, capabilities that were previously thought to be uniquely human.

The Development of AlphaGo

The development of **AlphaGo** began in **2014** at **DeepMind**, a London-based AI research lab founded by **Demis Hassabis, Shane Legg**, and **Mustafa Suleyman**. The team's goal was to create an AI capable of mastering complex tasks that required **decision-making** and **strategic planning**. Their early research had already demonstrated the power of **deep reinforcement learning**—a combination of **deep neural networks** and **reinforcement learning**—which had been successful in teaching AI to play Atari video games. Building on this foundation, DeepMind set its sights on Go, one of the most challenging games for human and machine intelligence alike.

AlphaGo was not the first Go-playing AI, but it was the first to integrate **deep learning** with **Monte Carlo Tree Search (MCTS)**. AlphaGo utilized two neural networks:

1. **The Policy Network**: This neural network evaluated the current board state and predicted the most promising moves. It was trained by **supervised learning**, using data from human professional games.

2. **The Value Network**: This network assessed the long-term prospects of a particular position on the

board and predicted the likelihood of winning from that position. It was trained by **reinforcement learning**, where AlphaGo played millions of games against itself to refine its understanding of good and bad positions.

By combining these two networks with MCTS, AlphaGo was able to explore a vast number of potential moves while focusing on the most promising lines of play. This approach allowed the AI to "think" more like a human player, making decisions based on **patterns** and **intuition** rather than brute-force calculations.

AlphaGo's ability to learn from both human games and its own experience was a revolutionary development in AI research. Unlike previous Go-playing programs that relied on predefined heuristics, AlphaGo's learning process enabled it to improve autonomously, continuously refining its strategies through self-play.

The Historic Match Against Lee Sedol

In March 2016, AlphaGo was pitted against **Lee Sedol**, one of the greatest Go players in the world, in a five-game match held in Seoul, South Korea. Prior to the match, many in the Go community were skeptical about AlphaGo's chances. Although AlphaGo had previously defeated **Fan Hui**, a

European Go champion, many believed that Sedol, with his creative and intuitive style, would easily outmatch the AI.

However, AlphaGo surprised everyone by winning the first three games, securing the series victory. The games demonstrated AlphaGo's ability to play not only strategically sound moves but also innovative and unexpected ones. In Game 2, for example, AlphaGo made a move that astonished both Sedol and Go commentators—a seemingly unconventional move that later proved to be a game-changer. This move was widely praised for its creativity, and it illustrated that AlphaGo was capable of playing moves that human players had never considered.

Lee Sedol managed to win Game 4, which remains a memorable moment in AI history. Sedol's victory was celebrated as a triumph of human intuition and creativity over machine learning. However, AlphaGo returned in Game 5 to secure a 4-1 victory, cementing its status as the strongest Go player in the world.

AlphaGo's victory over Lee Sedol was a defining moment for AI, showing that machines could not only mimic human cognition but, in some cases, outperform human experts in tasks that had long been considered beyond the reach of computers.

Technological Advancements: From Rule-Based AI to Reinforcement Learning

The success of AlphaGo marked a major leap in **AI capabilities**, moving beyond the traditional **rule-based systems** that had dominated AI research in earlier decades. Prior to AlphaGo, most successful game-playing AIs, such as **IBM's Deep Blue**, which defeated world chess champion **Garry Kasparov** in 1997, relied on brute-force search algorithms that evaluated millions of possible moves to determine the best one.

AlphaGo, on the other hand, represented a new paradigm in AI: **reinforcement learning** combined with **deep neural networks**. This approach allowed the AI to learn from experience and adapt its strategies over time, much like a human player. AlphaGo did not rely on pre-programmed rules or heuristics; instead, it learned through self-play, continually improving by playing millions of games against itself.

Deep reinforcement learning was a game-changer in AI research. By combining **deep learning**, which allows machines to recognize complex patterns, with **reinforcement learning**, which teaches machines to make decisions through trial and error, DeepMind was able to create an AI that could master a wide range of tasks without being explicitly programmed to do so. This approach was not limited to Go—it had applications in areas such as robotics, healthcare, and scientific research.

The development of AlphaGo also demonstrated the power of **transfer learning**—the ability of AI to apply knowledge gained in one domain (such as playing Go) to other, related domains. This breakthrough laid the groundwork for future AI systems, such as **AlphaZero**, which was capable of mastering not only Go but also chess and **shogi** (Japanese chess) using the same learning algorithms.

Cultural and Technological Impact

The cultural and technological impact of AlphaGo's victory over Lee Sedol cannot be overstated. In South Korea, where Go is a national pastime, the match was broadcast live on television, and millions of people followed the games online. AlphaGo's success sparked widespread interest in AI and its potential to reshape industries far beyond gaming.

In the world of Go, AlphaGo's victory changed the way the game was played. Professional Go players began to study AlphaGo's games to gain insights into new strategies and approaches. Some of AlphaGo's moves, which had initially been dismissed as unconventional, were later adopted by human players, who recognized their brilliance. In this sense, AlphaGo not only mastered the game but also contributed to the evolution of Go as a whole.

Beyond the gaming world, AlphaGo's success had profound implications for the future of AI. It demonstrated that machines could surpass human experts in tasks that required **creativity**, **intuition**, and **strategic thinking**—qualities that had long been considered uniquely human. This breakthrough raised important questions about the future of human-machine interaction and the role of AI in society.

AlphaGo's victory also ignited debates about the potential risks and ethical considerations surrounding AI. While AlphaGo's achievements were celebrated as a triumph of technology, they also raised concerns about the implications of AI for jobs, decision-making, and human creativity. As AI systems become more capable, there are growing concerns about their impact on the workforce and the need for ethical guidelines to ensure that AI is developed and used responsibly.

The Evolution of AlphaGo: AlphaGo Zero and AlphaZero

Following the success of AlphaGo, DeepMind continued to refine its AI systems, leading to the development of

AlphaGo Zero in 2017. Unlike its predecessor, AlphaGo Zero was not trained using human data; instead, it learned entirely through **self-play**, starting with no prior knowledge of Go and learning by playing millions of games against itself. Within just three days of training, AlphaGo Zero surpassed the original AlphaGo, achieving a level of play that was far beyond any human or machine player.

In 2018, DeepMind introduced **AlphaZero**, a more generalized version of the AI that could master multiple games, including chess, Go, and shogi, using the same algorithms. AlphaZero's ability to dominate multiple games demonstrated the versatility and power of **deep reinforcement learning**, marking another major leap toward the development of **artificial general intelligence (AGIAGI)**. AlphaZero's generalization capability reinforced DeepMind's vision of creating an AI that could learn to solve any problem across multiple domains, moving one step closer to achieving **artificial general intelligence**.

AlphaGo's development and subsequent victory over Lee Sedol in 2016 represented a defining moment in the history of **artificial intelligence**. It demonstrated that AI systems, through the use of **deep reinforcement learning** and **neural networks**, could surpass human experts in tasks requiring

strategic thinking, **intuition**, and **pattern recognition**. AlphaGo's success marked a major leap forward in AI research, pushing the boundaries of what machines could achieve and sparking new interest in the potential applications of AI beyond gaming.

The cultural impact of AlphaGo was equally significant, reshaping the way Go was played and studied while igniting debates about the future role of AI in society. AlphaGo's success inspired the development of more advanced systems like **AlphaGo Zero** and **AlphaZero**, demonstrating the versatility and power of **deep learning** in solving a wide range of complex problems.

In the broader context of AI research, AlphaGo served as a precursor to future innovations, such as **AlphaFold**, DeepMind's AI system that solved the **protein folding problem** and earned the company the **2024 Nobel Prize in Chemistry**. As we look ahead, the breakthroughs achieved by AlphaGo and its successors continue to shape the landscape of AI and raise important questions about the ethical and societal implications of increasingly powerful AI systems.

CHAPTER 4

AlphaFold: Revolutionizing Protein Structure Prediction

In 2020, **AlphaFold**, an AI system developed by **DeepMind**, made a breakthrough that solved one of the most fundamental challenges in **biology**—predicting the three-dimensional (3D) structure of proteins based solely on their amino acid sequences. This achievement addressed the so-called "**protein folding problem**," which had perplexed scientists for over five decades. The ability to predict protein structures with high accuracy is crucial for advancing our understanding of biological processes, and it holds immense potential for applications in **medicine**, **drug discovery**, and **biochemistry**. AlphaFold's success has not only transformed the field of **computational biology** but also opened new avenues for scientific discovery and innovation.

In this chapter, we will explore the development of AlphaFold, its significance in **protein structure prediction**, and the broader impact of its breakthrough on scientific research. We will also examine the role of **deep learning** in AlphaFold's success and how this AI-driven approach has changed the way we approach biological challenges.

The Protein Folding Problem: A 50-Year Challenge

Proteins are essential biological molecules that perform a wide variety of functions in living organisms, from **catalyzing chemical reactions** as enzymes to providing structural support in cells. These functions are determined by the protein's **3D shape**, which is a result of how the linear sequence of amino acids folds into a complex structure. Predicting how a protein will fold based on its amino acid sequence is known as the **protein folding problem**.

The challenge of protein folding lies in the sheer complexity of the process. Proteins consist of long chains of amino acids, and there are billions of possible ways these chains can fold into a 3D structure. While researchers have developed experimental techniques such as **X-ray crystallography** and **nuclear magnetic resonance (NMR) spectroscopy** to determine protein structures, these methods are time-consuming, expensive, and not applicable to all proteins.

For decades, scientists have sought to develop **computational models** that could predict protein structures accurately and quickly. However, the complexity of protein folding and the limitations of existing algorithms meant that accurate predictions were often elusive. The **Critical**

Assessment of Structure Prediction (CASP) competition, which has been held every two years since 1994, provided a benchmark for evaluating the performance of computational methods for predicting protein structures. Until AlphaFold, no computational system had come close to solving the protein folding problem with the accuracy required for real-world applications.

The Development of AlphaFold

The development of AlphaFold began at **DeepMind**, a company known for pushing the boundaries of **artificial intelligence (AI)**. Following their success with **AlphaGo**, the AI system that mastered the game of Go, DeepMind set its sights on a more ambitious challenge: solving the protein folding problem. By applying the same principles of **deep learning** that had been successful in game-playing AI, the DeepMind team aimed to develop an AI system that could predict protein structures with high accuracy.

AlphaFold was built using a **deep neural network** that was trained on a large dataset of known protein structures. The system works by taking a protein's amino acid sequence as input and using its neural network to predict the most likely 3D structure based on patterns it has learned from the training data. AlphaFold's architecture includes both **spatial**

and temporal dimensions, allowing it to predict how different parts of a protein interact and fold over time.

One of the key innovations in AlphaFold's design is its use of **attention mechanisms**—a technique originally developed for natural language processing (NLP)—to focus on the most relevant parts of the amino acid sequence and structure when making predictions. This approach allowed AlphaFold to model the complex interactions between different regions of a protein with greater accuracy than previous methods.

AlphaFold was first tested in the **CASP13 competition** in 2018, where it achieved impressive results, outperforming other computational methods in several categories. However, it was in the **CASP14 competition** in 2020 that AlphaFold truly demonstrated its potential. In CASP14, AlphaFold achieved a median score of **92.4 Global Distance Test (GDT)**, a metric used to measure the accuracy of predicted protein structures. This level of accuracy was comparable to that of experimental methods, and it represented a major breakthrough in the field of computational biology.

The Importance of Protein Structure Prediction

The ability to accurately predict protein structures based on their amino acid sequences has profound implications for **biological research** and **medicine**. Proteins are involved in nearly every biological process, and their structures are key to understanding their function. By knowing the 3D structure of a protein, scientists can gain insights into how it works, how it interacts with other molecules, and how mutations in the protein sequence may lead to diseases.

In **drug discovery**, for example, understanding protein structures is critical for designing drugs that can bind to specific proteins and modulate their activity. Many diseases, including cancer, Alzheimer's, and COVID-19, are caused by proteins that malfunction or misfold. By predicting the structure of these proteins, researchers can develop drugs that target them more effectively, leading to better treatments.

AlphaFold's ability to predict protein structures with high accuracy has the potential to accelerate the drug discovery process significantly. Traditionally, determining the structure of a protein using experimental methods could take years, but AlphaFold can generate predictions in a matter of hours or days. This speed allows researchers to focus their experimental efforts on validating the AI's predictions, saving time and resources.

In addition to drug discovery, protein structure prediction has applications in a wide range of fields, including **biochemistry**, **biotechnology**, and **synthetic biology**. For example, scientists can use AlphaFold to design **novel proteins** with specific functions, such as enzymes that can break down environmental pollutants or proteins that can be used in bioengineering applications.

Deep Learning and AlphaFold's Success

AlphaFold's success is largely due to the power of **deep learning**, a subset of machine learning that uses neural networks to model complex patterns in data. Deep learning has revolutionized many fields, from **natural language processing** to **computer vision**, by enabling AI systems to learn from large datasets and make predictions based on patterns they detect.

In the case of AlphaFold, deep learning allowed the AI to learn the intricate patterns of protein folding by analyzing vast amounts of data on known protein structures. The use of **convolutional neural networks (CNNs)**, a type of deep learning model particularly effective at processing grid-like data, was crucial for modeling the spatial relationships between different parts of a protein. Additionally, AlphaFold's use of **attention mechanisms** helped the system focus on the most relevant regions of the protein when making predictions.

One of the key challenges in protein structure prediction is capturing the **long-range dependencies** between amino acids that are far apart in the sequence but close together in the folded structure. AlphaFold's architecture was specifically designed to address this challenge by modeling both **local** and **global interactions** within the protein. This allowed the AI to make more accurate predictions about how different parts of the protein would come together in the final structure.

The use of deep learning in AlphaFold also enabled the system to improve over time through **self-supervised learning**. By training on a large dataset of known protein structures and their corresponding sequences, AlphaFold learned the rules of protein folding without being explicitly programmed with those rules. This ability to learn from data and generalize to new, unseen proteins is what sets AlphaFold apart from earlier computational methods that relied on handcrafted rules and heuristics.

Impact on Scientific Research in Biology

The breakthrough achieved by AlphaFold has already had a transformative impact on **scientific research** in **biology**. For decades, the lack of accurate and fast protein structure prediction tools had been a major bottleneck in fields such as **molecular biology**, **genetics**, and **biophysics**. AlphaFold has changed that by providing researchers with a tool that can predict protein structures with a level of accuracy comparable to experimental methods.

One of the most immediate impacts of AlphaFold has been in the field of **genomics**. With the rapid advancement of **genome sequencing technologies**, scientists now have access to vast amounts of genetic data. However, much of this data has been underutilized because researchers lacked the ability to determine the structures of the proteins encoded by these genes. AlphaFold has provided a solution to this problem by enabling researchers to predict the structures of proteins directly from their genetic sequences, unlocking new insights into how genes translate into function.

In addition to genomics, AlphaFold is also having a major impact on **structural biology**, a field that seeks to understand the relationship between the structure and function of biological molecules. By predicting the structures of proteins that are difficult or impossible to study experimentally, AlphaFold is enabling researchers to explore new areas of biology that were previously inaccessible. For example, AlphaFold has been used to predict the structures of proteins involved in **human diseases**, as well as proteins from **viruses** and **bacteria**, providing valuable information for developing new treatments and vaccines.

One of the most significant areas of impact has been in **COVID-19 research**. Early in the pandemic, AlphaFold was used to predict the structure of the **SARS-CoV-2 spike protein**, a key target for vaccines and therapies. These predictions helped researchers better understand how the virus infects cells and facilitated the development of vaccines, including the **mRNA vaccines** that have been instrumental in controlling the pandemic.

AlphaFold's success has also inspired new avenues of research in **protein engineering** and **synthetic biology**. By providing researchers with accurate predictions of protein structures, AlphaFold is enabling the design of **custom proteins** with specific functions, such as enzymes that can catalyze chemical reactions or proteins that can bind to specific molecules. This has the design of new enzymes and proteins, leading to advances in **biotechnology**,

environmental sustainability, and **agriculture**. For example, researchers are using AlphaFold to design enzymes that can degrade **plastic waste**, develop proteins that can improve the efficiency of biofuels, and create new crops that are more resistant to disease and environmental stress.

AlphaFold's Role in Personalized Medicine and Therapeutics

Another area where AlphaFold is expected to have a significant impact is in the development of **personalized medicine**. By predicting the structures of proteins that are implicated in genetic diseases, AlphaFold can help researchers design **targeted therapies** that are tailored to an individual's unique genetic makeup. This could lead to more effective treatments for diseases such as **cancer, genetic disorders**, and **neurodegenerative conditions**.

For example, many cancers are driven by mutations in proteins that cause them to misfold or malfunction. By predicting the structures of these mutant proteins, AlphaFold can help researchers design drugs that specifically target the defective proteins, preventing them from promoting tumor growth. This approach, known as **structure-based drug design**, has the potential to revolutionize the treatment of cancer and other diseases caused by protein misfolding.

AlphaFold's predictions are also being used to accelerate the development of **antibodies** and other **biologics**, which are increasingly being used to treat diseases such as **autoimmune disorders** and **infectious diseases**. By predicting how antibodies will bind to their target proteins, AlphaFold can help researchers design more effective and specific therapeutics, reducing the time and cost associated with drug development.

The Future of AlphaFold and AI in Biology

AlphaFold's success represents just the beginning of what is possible when **artificial intelligence** is applied to biological problems. As AI continues to advance, we can expect to see even greater breakthroughs in areas such as **protein design**, **genome editing**, and **synthetic biology**. In the future, AI systems like AlphaFold could be used to design entirely new proteins with functions that do not exist in nature, opening up new possibilities for **biotechnology** and **medicine**.

DeepMind has also made the decision to **open-source AlphaFold**, making the AI's predictions freely available to the global scientific community. This decision has already had a profound impact on research, with scientists around the world using AlphaFold's predictions to accelerate their work. By making this powerful tool accessible to everyone,

DeepMind has ensured that AlphaFold's benefits are shared widely, driving innovation across many fields of science.

Looking ahead, one of the key challenges for AlphaFold and other AI systems will be improving their ability to handle more complex biological systems. While AlphaFold has demonstrated remarkable success in predicting the structures of individual proteins, many biological processes involve interactions between multiple proteins and other molecules. Developing AI systems that can predict the structures of **protein complexes** and model their dynamic interactions in living cells will be a major focus of future research.

In addition to improving protein structure prediction, AI is likely to play a growing role in **genomics**, **systems biology**, and **molecular medicine**. By integrating AI with **high-throughput experimental techniques**, such as **CRISPR** and **single-cell sequencing**, researchers will be able to map out the intricate networks of genes, proteins, and pathways that underlie complex biological systems. This integrated approach will enable new discoveries in areas such as **developmental biology**, **cancer research**, and **neurobiology**, leading to new therapies and treatments for diseases that were previously untreatable.

AlphaFold represents a major breakthrough in the application of **artificial intelligence** to biological problems, solving a 50-year-old challenge that has long hindered progress in fields such as **medicine**, **drug discovery**, and **biochemistry**. By predicting the 3D structures of proteins with unprecedented accuracy, AlphaFold has opened new doors for scientific research and innovation, enabling researchers to better understand how proteins function and how they can be targeted for therapeutic intervention.

The use of **deep learning** in AlphaFold's development has been key to its success, allowing the AI to learn complex patterns in protein folding and make predictions based on large datasets of known protein structures. This approach has revolutionized the field of **computational biology** and set the stage for future advancements in areas such as **personalized medicine**, **protein engineering**, and **synthetic biology**.

As AI continues to advance, the impact of AlphaFold and similar systems will only grow, driving new discoveries in biology and beyond. Whether it's designing new drugs to treat diseases, creating sustainable solutions for environmental challenges, or engineering novel proteins for industrial applications, the potential for AI to transform science is vast. AlphaFold's success is a testament to the power of interdisciplinary collaboration, bringing together the fields of **AI**, **biology**, and **medicine** to solve some of the world's most pressing challenges.

In the coming years, AlphaFold and its successors will continue to shape the future of **biotechnology** and **biomedical research**, helping to unlock the mysteries of life at the molecular level and improving human health and well-being on a global scale. The journey of AlphaFold is just the beginning, and its legacy will inspire new generations of scientists and researchers to explore the potential of **artificial intelligence** in solving the most complex problems in biology.

CHAPTER 5

The 2024 Nobel Prize: Recognition of Breakthroughs in Science

In **2024**, the **Nobel Prize in Chemistry** was awarded to **Demis Hassabis**, **John Jumper**, and **David Baker** for their groundbreaking work in **protein structure prediction**, specifically for their development and application of **AlphaFold** and complementary techniques in the field of **computational biology**. This Nobel Prize marks a significant moment in the history of science, as it recognizes a **machine learning** and **artificial intelligence (AI)-driven** innovation for its profound impact on understanding the biological world. It underscores how **AlphaFold**—an AI system developed by **DeepMind**—has revolutionized the study of protein structures, solving a decades-old problem that has hindered progress in fields such as **biochemistry, medicine**, and **drug discovery**.

This chapter will delve into the significance of the **2024 Nobel Prize in Chemistry**, examining the scientific achievements of Hassabis, Jumper, and Baker, and how their work with **AlphaFold** has reshaped our understanding of biological systems. Furthermore, the chapter will explore the importance of this recognition for **artificial intelligence** and the broader implications of awarding a **Nobel Prize** to an AI-driven project, signaling a new era in which AI is increasingly recognized as a crucial tool in scientific discovery.

The 2024 Nobel Prize in Chemistry: Recognizing a Milestone in Protein Structure Prediction

The **Nobel Prize** is one of the highest honors in the scientific community, awarded annually to individuals or groups whose work has profoundly advanced humanity's understanding of the natural world. In **2024**, the Nobel Prize in Chemistry was awarded to **Demis Hassabis**, **John Jumper**, and **David Baker** for their contributions to solving the **protein folding problem**—one of the most fundamental and challenging problems in biology. Their efforts culminated in the development of **AlphaFold**, an AI system capable of predicting the three-dimensional structure of proteins based on their amino acid sequences with unprecedented accuracy.

Proteins are the molecular machines that carry out almost every function in a living organism, from catalyzing biochemical reactions to forming the structural components of cells. The function of a protein is directly related to its shape, which is determined by how the chain of amino acids folds into a 3D structure. Understanding protein structures is critical for fields such as **biochemistry**, **molecular biology**, and **medicine**, but experimentally determining these structures is a laborious and expensive process that can take years. The ability to accurately predict protein structures computationally has long been a goal of researchers because it would accelerate the study of biological processes and enable faster drug discovery.

For decades, the **protein folding problem** resisted resolution. Experimental methods such as **X-ray crystallography** and **nuclear magnetic resonance (NMR)** have been used to determine protein structures, but they are time-consuming, expensive, and not applicable to all proteins. Computational methods offered a potential solution, but previous attempts to develop algorithms capable of accurately predicting protein structures had limited success. This is where **AlphaFold** comes in: by leveraging **deep learning** and **neural networks**, AlphaFold has revolutionized the field of **computational biology**, achieving a level of accuracy that rivals experimental methods.

The Role of AlphaFold in Protein Structure Prediction

The core breakthrough behind AlphaFold's success is its ability to predict the **3D structure** of a protein based solely on its amino acid sequence. This is a monumental achievement because it allows researchers to bypass the costly and time-consuming experimental methods traditionally used to study proteins. AlphaFold's predictions are based on patterns learned from large datasets of known protein structures. By training a deep neural network on these datasets, AlphaFold was able to learn the complex rules that govern how a linear chain of amino acids folds into a functional 3D structure.

At the heart of AlphaFold's success is the use of **attention mechanisms**—a machine learning technique originally developed for natural language processing—to focus on the most relevant parts of a protein sequence when making predictions. This enables AlphaFold to accurately predict the interactions between distant regions of a protein sequence, which are crucial for determining its final structure. AlphaFold's architecture also includes a **spatial graph representation** of protein structures, allowing it to model the geometric relationships between atoms within the protein.

AlphaFold's ability to predict protein structures with near-experimental accuracy has far-reaching implications for **biology** and **medicine**. Proteins play a central role in almost every biological process, and understanding their structure is key to developing new **therapeutics**. For example, many diseases, including cancer and neurodegenerative disorders, are caused by proteins that misfold or malfunction. AlphaFold's predictions can help researchers understand how these proteins work and design drugs that target them more effectively.

In the **Critical Assessment of Structure Prediction (CASP14)** competition in 2020, AlphaFold achieved a **median Global Distance Test (GDT)** score of **92.4**, a level of accuracy that was comparable to experimental methods. This achievement marked a turning point in the field of protein structure prediction, as it demonstrated that AI could solve a problem that had challenged biologists for decades. The Nobel Prize in Chemistry in 2024 recognized the significance of this breakthrough and its potential to transform many areas of science.

The Contributions of Demis Hassabis, John Jumper, and David Baker

The success of AlphaFold is the result of a collaborative effort by some of the world's leading researchers in **artificial intelligence** and **biochemistry**. The 2024 Nobel Prize in Chemistry was awarded to three individuals whose contributions were instrumental in the development of this groundbreaking AI system:

- **Demis Hassabis**, co-founder and CEO of **DeepMind**, played a leading role in the overall development of AlphaFold and its application to solving the protein folding problem. With a background in **neuroscience** and **computer science**, Hassabis brought a unique interdisciplinary perspective to the project, combining insights from human cognition with cutting-edge AI techniques.

- **John Jumper**, a senior researcher at DeepMind, was the technical lead on the AlphaFold project. His expertise in **machine learning** and **biophysics** was crucial for designing the deep neural networks that underpin AlphaFold's ability to predict protein structures. Jumper's work helped refine AlphaFold's architecture and improve its accuracy, making it one of the most powerful tools in computational biology.

- **David Baker**, a biochemist and professor at the University of Washington, was recognized for his complementary work on **protein design** and the development of the **Rosetta** software, a computational tool used to model protein structures. Baker's research on **de novo protein design**—the creation of entirely new proteins not found in nature—has been enabled by advances in computational biology like AlphaFold.

Together, these three scientists have reshaped our understanding of how proteins fold and interact, paving the way for new discoveries in **biomedicine** and **biotechnology**.

The Broader Impact of AlphaFold on Science and Medicine

AlphaFold's success has far-reaching implications for the scientific community, particularly in the fields of **medicine**, **biochemistry**, and **drug discovery**. By providing accurate predictions of protein structures, AlphaFold is helping researchers accelerate the development of new drugs and treatments for diseases caused by malfunctioning proteins. For example, in **cancer research**, AlphaFold has been used to study the structures of **oncoproteins**—proteins that drive tumor growth—enabling researchers to design drugs that specifically target these proteins and inhibit their activity.

One of the most significant applications of AlphaFold has been in the study of **viral proteins**, particularly those involved in infectious diseases such as **COVID-19**. Early in the pandemic, AlphaFold was used to predict the structure of the **SARS-CoV-2 spike protein**, a critical component of the virus that enables it to enter human cells. This information was invaluable for the development of **vaccines** and **therapeutics** that target the spike protein, including the **mRNA vaccines** that have played a central role in controlling the pandemic.

In addition to its applications in drug discovery, AlphaFold is also being used to advance our understanding of **genetics** and **molecular biology**. By predicting the structures of proteins encoded by the human genome, AlphaFold is helping researchers identify new **drug targets** and develop treatments for genetic diseases. The ability to predict protein structures from genomic data has the potential to revolutionize the field of **personalized medicine**, enabling doctors to tailor treatments to an individual's genetic profile.

Beyond medicine, AlphaFold's predictions are being used in fields such as **environmental science** and **biotechnology**. For example, researchers are using AlphaFold to design **enzymes** that can break down **plastic waste**, develop **biofuels** that are more efficient and sustainable, and engineer crops that are more resistant to **disease** and **climate change**. The ability to design new proteins with specific functions has opened up new possibilities for solving some of the world's most pressing challenges.

AlphaFold and the Future of AI in Scientific Discovery

The award of the **2024 Nobel Prize in Chemistry** to an AI-driven project like AlphaFold represents a major shift in how **artificial intelligence** is recognized in the scientific community. Traditionally, the Nobel Prize has been awarded to individuals who have made significant contributions to fields such as **physics**, **chemistry**, and **medicine** through experimental or theoretical research. The recognition of an AI system like AlphaFold marks a new era in which AI is increasingly seen as a valuable tool for scientific discovery, capable of solving complex problems that were previously thought to be beyond the reach of machines.

AlphaFold's success has demonstrated the power of **deep learning** and **neural networks** to tackle complex tasks that have previously eluded traditional scientific approaches. The success of **AlphaFold** in protein structure prediction shows how AI can augment human intelligence and push the boundaries of what is possible in science.

The impact of AlphaFold extends beyond the immediate applications in **medicine** and **biotechnology**. It represents a broader shift in the way scientists approach problems that involve complex systems and vast amounts of data. AI systems like AlphaFold are enabling researchers to solve problems that were previously considered intractable, accelerating the pace of scientific discovery and opening up new frontiers in fields such as **quantum chemistry**, **materials science**, and **systems biology**.

As AI continues to advance, its role in scientific discovery is likely to grow even more significant. The development of AI systems that can analyze large datasets, identify patterns, and make predictions is transforming how researchers work, allowing them to tackle problems that were once thought to be beyond human understanding. In the coming years, we can expect to see AI play an increasingly important role in areas such as **genomics**, **drug design**, and **climate science**, helping researchers address some of the most pressing challenges facing humanity.

The awarding of the **2024 Nobel Prize in Chemistry** to **Demis Hassabis, John Jumper,** and **David Baker** for their work on **protein structure prediction** marks a major milestone in the history of science. It recognizes the profound impact of **AlphaFold,** an AI system that has revolutionized the field of **computational biology** and opened new possibilities for scientific discovery. By solving the **protein folding problem,** AlphaFold has transformed our understanding of biological systems and accelerated the development of new **therapies** for diseases, as well as **biotechnological solutions** for global challenges.

The recognition of AlphaFold with a Nobel Prize also signals a shift in how **artificial intelligence** is viewed within the scientific community. AI is no longer seen as a tool for automating tasks or solving specific problems; it is now recognized as a fundamental driver of scientific innovation, capable of solving some of the most complex and challenging problems in science. The success of AlphaFold is a testament to the power of **AI** and its potential to transform fields as diverse as **medicine, biotechnology,** and **environmental science.**

CHAPTER 6
The Broader Implications of AI in Science and Medicine

The field of **artificial intelligence (AI)** has long been associated with the development of technologies like **speech recognition, self-driving cars**, and **personal assistants**. However, in recent years, AI has made transformative strides in areas such as **science** and **medicine**, proving its potential to revolutionize these fields. The groundbreaking success of **AlphaFold**, developed by **DeepMind**, in predicting **protein structures** has demonstrated how AI can accelerate research and discovery, solving problems that were previously insurmountable due to the limitations of human labor and traditional computational methods.

AlphaFold, alongside similar AI-driven technologies, has the potential to reshape numerous sectors, from **drug discovery** to **vaccine development** and **industrial biotechnology**. These systems are enabling researchers to unlock biological mysteries at a faster pace and with greater accuracy than ever before, leading to innovations in **personalized medicine, environmental sustainability**, and even **synthetic biology**.

This chapter will explore the broader implications of AI in science and medicine, looking at how AI systems like AlphaFold are being used to accelerate research, how they impact **drug discovery** and **vaccine development**, and how they contribute to innovations in **industrial enzyme design** and environmental sustainability.

AI-Driven Acceleration of Scientific Research

AlphaFold has emerged as a game-changer in the field of **biological sciences**, particularly in **protein structure prediction**. Proteins, as the workhorses of biological systems, carry out essential functions ranging from catalyzing reactions as enzymes to facilitating cellular communication and maintaining structural integrity. Understanding the **three-dimensional (3D) structure** of a protein is critical to understanding its function. However, determining these structures through traditional experimental methods like **X-ray crystallography**, **nuclear magnetic resonance (NMR)**, and **cryo-electron microscopy** is a time-consuming and expensive process.

AlphaFold, using **deep learning** algorithms, solves this challenge by predicting a protein's 3D structure based solely on its amino acid sequence. This not only accelerates the pace of research but also allows scientists to study proteins that are difficult or impossible to analyze using experimental techniques. For example, AlphaFold has been used to predict the structures of previously unknown proteins, providing researchers with insights into how these proteins function in biological processes and how they might be targeted for therapeutic intervention.

Deep learning, the AI technique behind AlphaFold, has proven highly effective in fields where massive datasets exist, allowing AI models to identify patterns and predict outcomes with remarkable accuracy. In biology, AlphaFold's model was trained on thousands of known protein structures, enabling it to make highly accurate predictions for new proteins. The availability of these predictions has had a profound impact on research across fields such as **biochemistry**, **molecular biology**, **genetics**, and **biomedical engineering**.

Impact on Drug Discovery

Perhaps the most significant application of AI systems like AlphaFold is in the field of **drug discovery**. Traditionally, drug development has been an arduous process, taking years or even decades to progress from initial research to an approved therapy. One of the major bottlenecks in this process is the identification of **drug targets**—the proteins within the body that drugs interact with to elicit a therapeutic effect. Understanding the structure of these proteins is critical for designing drugs that can bind to them effectively.

AlphaFold addresses this challenge by providing accurate and rapid predictions of protein structures, enabling researchers to identify potential drug targets much more efficiently. By knowing the precise structure of a protein, drug developers can design **small molecules** or **biologics** that interact with the target protein in a specific way, increasing the likelihood of success in drug development. This is particularly important for diseases caused by **mutations** or **misfolding** of proteins, such as **cancer**, **Alzheimer's disease**, and **cystic fibrosis**.

Structure-based drug design is a process in which scientists use the 3D structure of a target protein to design molecules that fit into its active site or other important regions. AlphaFold's ability to predict these structures has greatly enhanced this process, reducing the time and cost involved in developing new therapies. For example, researchers working on **cancer therapeutics** have used AlphaFold to predict the structures of **oncoproteins**, which are proteins that drive tumor growth when mutated or overexpressed. By targeting these proteins with specifically designed drugs, scientists can potentially inhibit tumor growth more effectively.

In addition to accelerating the early stages of drug discovery, AlphaFold has applications in later stages, such as **lead optimization**. Once a promising drug candidate has been identified, AlphaFold can be used to predict how small modifications to the drug's chemical structure will affect its binding to the target protein, enabling researchers to fine-tune the drug's efficacy and reduce its potential side effects.

The role of AI in drug discovery is not limited to AlphaFold. Other AI-driven platforms are being used to analyze **biological data**, **predict drug interactions**, and even design entirely new molecules. Together, these systems are transforming the way drugs are developed, offering the potential to shorten development timelines and bring new treatments to patients more quickly.

Applications in Vaccine Development

Another area where AlphaFold has made a significant impact is in **vaccine development**. Vaccines typically work by stimulating the immune system to recognize and neutralize pathogens such as **viruses** and **bacteria**. This process relies on the immune system's ability to recognize **antigens**, which are specific proteins found on the surface of pathogens. Understanding the structure of these antigens is critical for designing vaccines that elicit a strong immune response.

During the **COVID-19 pandemic**, AlphaFold was used to predict the structure of the **SARS-CoV-2 spike protein**, a key antigen that the virus uses to enter human cells. This information was instrumental in the development of vaccines, including **mRNA vaccines** like those produced by **Pfizer-BioNTech** and **Moderna**. By providing accurate structural information about the spike protein, AlphaFold helped researchers design vaccines that targeted this protein effectively, leading to the rapid development of safe and effective vaccines.

The ability to predict **viral protein structures** quickly and accurately has the potential to revolutionize vaccine development for other diseases as well. For example, AlphaFold could be used to predict the structures of proteins from viruses that cause diseases such as **influenza**, **HIV**, or **Zika**, enabling researchers to design vaccines that target these pathogens more precisely. In addition to traditional vaccines, AI-driven systems like AlphaFold can be used to develop **therapeutic vaccines**, which are designed to treat existing infections or chronic diseases such as cancer by stimulating the immune system to target specific proteins expressed by cancer cells or pathogens.

Moreover, AI is being used to optimize vaccine formulations by predicting how different components of a vaccine will interact with the immune system. This allows researchers to design vaccines that produce stronger and longer-lasting immune responses, improving their efficacy in preventing disease.

Enzyme Design for Industrial Purposes

In addition to its applications in medicine, AlphaFold and similar AI systems are being used to design **enzymes** for **industrial purposes**. Enzymes are proteins that catalyze chemical reactions, and they are used in a wide range of industries, from **food production** and **biofuels** to **pharmaceuticals** and **environmental cleanup**. Designing new enzymes with specific properties can lead to more efficient and sustainable industrial processes.

For example, in the field of **biofuels**, researchers are using AlphaFold to design enzymes that can break down **cellulose** from plant biomass into **sugars**, which can then be fermented into biofuels such as **ethanol**. By optimizing the structure of these enzymes, scientists can improve their efficiency and reduce the cost of biofuel production, making it a more viable alternative to fossil fuels.

In the **food industry**, enzymes are used to improve the texture, flavor, and shelf life of products. AlphaFold is helping researchers design enzymes that can be used in the production of **dairy alternatives**, **gluten-free products**, and other specialty foods, offering new options for consumers with dietary restrictions. For example, enzymes that break down **lactose** in milk are essential for producing **lactose-free dairy products**, and enzymes that degrade **gluten** can be used to create gluten-free breads and pastries.

One of the most promising applications of enzyme design is in the field of **environmental sustainability**. Researchers are using AlphaFold to design enzymes that can break down **plastic waste** and other environmental pollutants. For example, enzymes that degrade **polyethylene terephthalate (PET)**, a common plastic used in bottles and packaging, are being engineered to break down PET into its basic components, which can then be recycled into new plastic products. This approach has the potential to reduce the environmental impact of plastic waste, which is a major global problem.

AI-driven enzyme design is also being used in the **pharmaceutical** and **chemical industries** to develop enzymes that catalyze specific reactions, reducing the need for toxic chemicals and energy-intensive processes. By making industrial processes more efficient and sustainable, AI systems like AlphaFold are contributing to the development of a **circular economy**, where materials are reused and waste is minimized.

The Environmental Impact of AI-Designed Proteins

The broader environmental implications of AI-designed proteins go beyond enzyme design. As climate change and environmental degradation continue to pose major challenges, AI has the potential to offer innovative solutions. In **agriculture**, for example, AI systems like AlphaFold are being used to design proteins that can improve crop resilience to **drought**, **pests**, and **diseases**. By engineering crops that require fewer chemical inputs and are more resistant to environmental stressors, researchers can reduce the environmental impact of climate change, reduce the need for **fertilizers** and **pesticides**, which contribute to environmental degradation. For example, **AI-designed proteins** can be used to enhance the **nitrogen-fixing ability** of certain plants, reducing the need for synthetic fertilizers that often lead to **water pollution** and the release of **greenhouse gases**.

Moreover, AI-driven innovations in **biomaterials** are making it possible to design **sustainable alternatives** to traditional materials like plastics and chemicals. For instance, researchers are using AI to design proteins that can be used in the production of **biodegradable materials**, which break down more easily in the environment compared to traditional plastics. These materials can be used in packaging, textiles, and construction, offering a sustainable alternative to petroleum-based products.

In the energy sector, AI-designed proteins are being explored for their potential to improve **solar energy capture** and **storage**. Proteins that can facilitate the efficient conversion of **solar energy** into **chemical energy** (such as hydrogen) are being designed to help drive the shift toward renewable energy sources. These proteins could be used in **artificial photosynthesis** systems, which mimic the natural process by which plants convert sunlight into energy, offering a sustainable solution to the world's growing energy needs.

AI's contribution to **environmental sustainability** also extends to **carbon capture**. Proteins designed by AI are being explored for their ability to capture and convert carbon dioxide (CO2), a major greenhouse gas, into useful products. By using these proteins in **carbon capture and storage (CCS)** systems, it may be possible to reduce the amount of CO2 in the atmosphere, mitigating the effects of climate change.

AI as a Tool for Solving Complex Problems

One of the broader implications of AI systems like AlphaFold is their ability to tackle **complex problems** that involve massive amounts of data and intricate relationships between variables. In many ways, AI has become a powerful tool for **augmenting human intelligence**, allowing researchers to solve problems that were previously considered intractable.

In **systems biology**, for example, AI is being used to model complex biological systems, such as the **networks of genes**, **proteins**, and **metabolites** that govern cellular behavior. By using AI to analyze these systems, researchers can gain insights into how diseases arise and how they can be treated more effectively. This approach is particularly valuable in understanding **multifactorial diseases**, such as cancer and autoimmune disorders, which involve the interplay of multiple genetic and environmental factors.

In **precision medicine**, AI is enabling the development of **personalized treatments** that are tailored to an individual's unique genetic makeup. By analyzing large datasets of genomic, proteomic, and clinical information, AI systems can identify the genetic mutations and biological pathways that contribute to disease in a particular patient. This information can then be used to design targeted therapies that are more effective and have fewer side effects than traditional treatments.

In **quantum chemistry**, AI is being used to model the behavior of **molecules** and **chemical reactions** at the quantum level. This approach has the potential to accelerate the discovery of new materials, drugs, and catalysts by enabling researchers to explore chemical space more efficiently. AI-driven innovations in quantum chemistry could lead to breakthroughs in fields ranging from **energy storage** to **materials science**.

The broader implications of AI in science and medicine are profound, as demonstrated by the success of AlphaFold and similar systems. These AI technologies are revolutionizing fields such as **drug discovery**, **vaccine development**, **enzyme design**, and **environmental sustainability**, offering new tools for solving some of the world's most pressing challenges. By accelerating research and providing new insights into the structure and function of biological systems, AI is enabling scientists to develop more effective treatments for diseases, design sustainable materials, and create solutions for global environmental problems.

As AI continues to evolve, its impact on science and medicine will only grow, offering new possibilities for discovery and innovation. From **precision medicine** to **renewable energy**, AI-driven solutions are poised to transform the way we approach the most complex and urgent problems facing humanity. The recognition of AI's potential by institutions such as the **Nobel Prize committee**, which awarded the **2024 Nobel Prize in Chemistry** to the team behind AlphaFold, marks a significant step toward integrating AI into the heart of scientific discovery. This new era of AI-enhanced research promises to unlock new frontiers in science and technology, leading to a future where AI and human ingenuity work together to solve the world's most challenging problems.

CHAPTER 7
AI Ethics and the Future of Machine Learning

The rapid advancements in **artificial intelligence (AI)**, particularly through breakthroughs like **AlphaFold** and other innovations at **DeepMind**, have opened up new frontiers in science, medicine, and technology. However, alongside these exciting developments comes an urgent need to address the **ethical implications** of AI. As AI technologies evolve, so do the challenges related to their responsible use, ensuring they do not inadvertently cause harm or perpetuate biases in society. **Demis Hassabis**, the co-founder and CEO of **DeepMind**, has been a leading voice in the conversation on **AI ethics**, advocating for a careful and principled approach to AI development. His perspective highlights both the promise of AI and the responsibilities researchers have in ensuring these technologies benefit humanity while minimizing potential risks.

This chapter will explore Demis Hassabis' views on AI ethics, the ethical challenges posed by advanced AI technologies, including issues such as **job displacement**, **bias**, and the risk of **misuse**, and the ongoing efforts to promote **responsible AI development**. By examining these themes, we can better understand the ethical landscape of AI and what the future may hold for machine learning in both scientific and societal contexts.

Demis Hassabis' Views on AI Ethics: Balancing Innovation and Responsibility

Demis Hassabis has been at the forefront of discussions surrounding the ethical dimensions of AI. His perspective, shaped by his work at **DeepMind** and his broader vision of AI's potential, emphasizes the need for **transparency**, **accountability**, and a **cautious approach** to the deployment of AI systems. While Hassabis acknowledges the transformative power of AI, he also believes that researchers and developers have a responsibility to consider the societal impact of the technologies they create.

One of the key ethical principles championed by Hassabis is the importance of **human oversight** in AI systems. He has consistently argued that, regardless of how advanced AI becomes, there should always be a clear understanding of how these systems make decisions, and human beings should remain accountable for their actions. This is particularly important in areas like **healthcare**, **finance**, and

law enforcement, where the consequences of AI-driven decisions can have significant and far-reaching effects.

Hassabis has also spoken about the importance of **fairness** and **equity** in AI. He believes that AI should be developed in a way that **benefits all of humanity**, not just a select few. This includes ensuring that AI systems are designed to reduce **bias** rather than exacerbate it and that their deployment does not disproportionately affect marginalized or vulnerable communities. In a world where AI has the potential to automate tasks, make predictions, and even influence public policy, ensuring fairness is paramount to avoid **perpetuating existing inequalities**.

The Ethical Challenges of Advanced AI Technologies

As AI technologies continue to advance, several key ethical challenges have emerged. These include concerns about **job displacement**, **bias in AI systems**, and the **potential for misuse**, particularly in areas such as **surveillance** and **military applications**. These challenges require careful consideration and proactive measures to ensure that the benefits of AI are maximized while minimizing harm.

1. **Job Displacement**: One of the most widely discussed ethical concerns surrounding AI is the potential for **job displacement**. As AI systems

become more capable of automating tasks across industries, from manufacturing to customer service, there is a fear that large segments of the workforce could be displaced by machines. While automation has historically led to shifts in the labor market—often creating new jobs to replace those that are lost—the scale and speed of AI-driven automation could create significant economic and social disruption. This is particularly true in industries where tasks are repetitive and predictable, making them prime candidates for automation.

Demis Hassabis and other AI leaders have acknowledged the need to address this challenge proactively. One possible solution is **reskilling** and **upskilling** workers whose jobs may be threatened by AI. Governments and private sectors must collaborate to provide education and training programs that equip workers with the skills they need to succeed in a more AI-driven economy. Additionally, **universal basic income (UBI)** and other social safety nets have been proposed as ways to support individuals who may be displaced by automation.

2. **Bias in AI Systems**: Another significant ethical concern in AI development is the issue of **bias**. AI

systems are only as good as the data they are trained on, and if the training data contains biases—whether based on **race**, **gender**, **socioeconomic status**, or other factors—those biases can be reflected and even amplified in the AI's predictions and decisions. This has been particularly problematic in areas such as **criminal justice**, **hiring**, and **credit scoring**, where biased AI systems can lead to **unfair outcomes** and **discrimination**.

Hassabis has emphasized the importance of addressing **algorithmic bias** by ensuring that AI systems are trained on diverse and representative datasets. Additionally, researchers must be transparent about the limitations of their models and regularly audit AI systems for unintended biases. DeepMind, under Hassabis' leadership, has invested in research on **ethical AI** and **fairness** in machine learning, recognizing that mitigating bias is critical to building trust in AI systems.

3. **Surveillance and Military Applications**: The potential for AI to be misused in areas such as **surveillance** and **military operations** is another pressing ethical issue. AI-powered surveillance technologies, such as **facial recognition**, raise concerns about **privacy** and **civil liberties**. In some

cases, these technologies have been used by governments to monitor citizens without their consent, leading to fears of a **surveillance state**. Similarly, the use of AI in **military applications**, particularly in **autonomous weapons systems**, has sparked debates about the morality of allowing machines to make decisions about life and death.

Hassabis and DeepMind have taken a strong stance against the development of **autonomous weapons**, joining other AI researchers in calling for an international ban on so-called "**killer robots**." In 2018, DeepMind was one of the signatories to a pledge stating that they would not participate in the development of AI systems intended for autonomous weaponry. This reflects Hassabis' belief that AI should be used for **peaceful and constructive purposes**, not for exacerbating conflicts or enabling state surveillance.

Promoting Responsible AI Development and Transparency in Research

Demis Hassabis and DeepMind have made significant efforts to promote **responsible AI development**, recognizing that AI's impact on society will depend on how it is developed and deployed. One of the ways Hassabis has sought to ensure that AI is developed responsibly is through the establishment of **ethical guidelines** and **governance structures** within DeepMind. This includes the creation of an **AI Ethics Board**, which oversees the ethical implications of DeepMind's research and ensures that the company's technologies are used in ways that align with their ethical principles.

In addition to internal governance, Hassabis has also been an advocate for **transparency** in AI research. Transparency is critical for building **trust** in AI systems, as it allows the public and other stakeholders to understand how these systems work and how decisions are made. One of the key challenges in AI is the issue of **black box decision-making**, where even the developers of AI systems may not fully understand how a system arrived at a particular conclusion. To address this, Hassabis has called for greater transparency in AI development, including the need to explain AI models in ways that are understandable to non-experts.

DeepMind's commitment to transparency was demonstrated when the company made the **AlphaFold** protein structure predictions freely available to the global scientific community. By open-sourcing the **AlphaFold Protein Structure Database**, DeepMind enabled researchers around the world to access one of the most powerful tools in biology, accelerating scientific discovery in fields such as **medicine**, **biochemistry**, and **genomics**. This decision not only benefited the scientific community but also helped build trust in AI's potential to contribute to the common good.

Moreover, Hassabis has emphasized the importance of **multidisciplinary collaboration** in AI development. He believes that the future of AI will require input from experts in **ethics**, **law**, **philosophy**, and **social sciences**, in addition to computer scientists and engineers. This interdisciplinary approach ensures that AI systems are developed in ways that take into account the broader societal and ethical implications of their use.

The Future of Machine Learning: Toward Ethical and Equitable AI

As AI continues to advance, its role in shaping the future of society will only become more significant. The ethical challenges that accompany these advancements will require ongoing dialogue, research, and policy development to ensure that AI is used in ways that **promote human flourishing** and avoid harm.

One of the key challenges for the future of machine learning is the need to develop AI systems that are **accountable** and **transparent**. This will require advancements in **explainable**

AI (XAI), a subfield of AI focused on making machine learning models more interpretable and understandable to humans. By ensuring that AI systems are not opaque black boxes, researchers can build trust in these technologies and ensure that they are used responsibly.

Another important aspect of the future of AI is the need to address the **ethical dilemmas** posed by autonomous systems. As AI systems become more capable of making decisions autonomously—whether in **self-driving cars**, **robotics**, or healthcare—there will be a need to ensure that these systems adhere to **ethical guidelines** and that humans remain accountable for their actions. This will require ongoing collaboration between AI researchers, policymakers, and ethicists to develop **regulatory frameworks** that ensure AI is used for the benefit of society.

Finally, the future of machine learning will require addressing issues of **inequality** and **access**. While AIto ensure everyone benefits from AI's advancements, there is a need to address **inequality** in access to AI tools and technologies. The benefits of AI should not be limited to wealthy nations or organizations; instead, efforts must be made to democratize AI and ensure that its benefits are shared globally. This includes providing access to AI education and resources in **low-income countries** and **underrepresented communities** to prevent the widening of existing social and economic disparities.

The integration of AI into society is an ongoing journey that requires balancing innovation with responsibility. Demis Hassabis' leadership in AI ethics, his commitment to transparency, and his focus on the responsible use of technology are paving the way for an ethical and equitable future of machine learning. As AI continues to evolve, it will be crucial for researchers, policymakers, and society to work together to harness its potential while mitigating its risks, ensuring that AI remains a force for good in the world.

The rapid advancements in AI, particularly through projects like AlphaFold, have demonstrated the immense potential of machine learning in solving complex scientific and societal challenges. However, as these technologies become more powerful, the ethical implications of their use must be carefully considered. Demis Hassabis' work at DeepMind, along with his advocacy for transparency, fairness, and responsible AI development, highlights the importance of taking a **principled approach** to AI research.

The ethical challenges surrounding AI—ranging from job displacement to bias and the potential for misuse in surveillance and warfare—require proactive measures and collaboration across disciplines. By fostering a culture of **accountability** and **transparency**, AI developers can ensure that these technologies benefit humanity while minimizing harm. The future of machine learning lies in its ability to balance innovation with ethical considerations, and with leaders like Hassabis at the helm, the future of AI looks promising, provided we address its ethical challenges head-on.

CHAPTER 8
Collaborations with David Baker and John Jumper

The groundbreaking success of **AlphaFold** in solving the **protein structure prediction problem** was not a solitary achievement but the result of numerous collaborations among some of the world's most brilliant minds in **artificial intelligence (AI)** and **biochemistry**. Among these collaborations, the partnership between **Demis Hassabis**, **John Jumper**, and **David Baker** stands out as a pivotal force in advancing **biotechnological innovations**. Their combined efforts, culminating in the 2024 **Nobel Prize in Chemistry**, have reshaped the way scientists approach protein folding, offering new solutions in **medicine**, **drug discovery**, and **industrial applications**.

This chapter delves into the significance of the collaboration between Hassabis, Baker, and Jumper, focusing on how their joint efforts accelerated advancements in **protein structure prediction**. We will explore Baker's contributions through the **Rosetta** software and how it complements AlphaFold's AI-driven approach. Furthermore, we will discuss how these collaborations are shaping the future of **biotechnology**, influencing fields as diverse as **personalized medicine**, **vaccine development**, and **environmental sustainability**.

Demis Hassabis and AlphaFold: Revolutionizing Protein Folding

The foundation of this collaboration lies in the revolutionary achievements of **AlphaFold**, developed by **DeepMind** under the leadership of **Demis Hassabis** and **John Jumper**. Protein folding has long been one of the most perplexing challenges in **molecular biology**. Proteins are complex molecules made of **amino acids** that fold into precise three-dimensional structures, which determine their function. Misfolded proteins are often associated with diseases, including **cancer**, **Alzheimer's**, and **cystic fibrosis**.

Traditional experimental methods for determining protein structures—such as **X-ray crystallography**, **nuclear magnetic resonance (NMR)**, and **cryo-electron microscopy**—are time-consuming and expensive. For decades, researchers have sought to develop computational tools that could accurately predict protein structures based on amino acid sequences, but previous attempts were limited in their accuracy. AlphaFold, with its ability to predict protein structures with near-experimental accuracy, was a game-changer.

AlphaFold utilizes **deep learning** and **neural networks** to predict the 3D structure of a protein. The system was trained on tens of thousands of known protein structures, allowing it to learn the rules of protein folding. In 2020, AlphaFold achieved a **92.4 Global Distance Test (GDT)** score in the **CASP14** competition, a level of accuracy that rivaled experimental methods. This achievement not only marked a breakthrough in computational biology but also paved the way for further innovations in **drug discovery**, **genetics**, and **biochemistry**.

David Baker and the Rosetta Software: A Complementary Approach

While AlphaFold represents a breakthrough in AI-driven protein structure prediction, it is important to recognize the complementary contributions of **David Baker** and his work on the **Rosetta** software. Developed at the **University of Washington**, Rosetta is a powerful tool used for **protein modeling** and **structure prediction**. Unlike AlphaFold, which relies on **deep learning**, Rosetta is based on **physics-based simulations** that predict how proteins fold by evaluating the interactions between atoms and molecules.

Rosetta has been widely used by researchers to model protein structures, design novel proteins, and predict the effects of **mutations** on protein function. One of the key strengths of Rosetta is its versatility. It can be applied to a wide range of problems in structural biology, from predicting the structures of **antibodies** to designing **enzymes** for industrial applications. For example, Baker's lab used Rosetta to design proteins that could neutralize the **SARS-CoV-2 virus**, contributing to the development of therapeutics during the **COVID-19 pandemic**.

Baker's work on Rosetta complements AlphaFold by providing researchers with additional tools for studying protein folding. While AlphaFold excels at predicting the structure of a protein based on its amino acid sequence, Rosetta's **de novo protein design** capabilities allow scientists to design entirely new proteins with specific functions. This is particularly useful in the development of **biologics**, **vaccines**, and **industrial enzymes**. By combining the strengths of both systems, researchers can achieve a more comprehensive understanding of protein structures and their applications.

John Jumper's Role in AlphaFold's Development

John Jumper, a senior researcher at DeepMind, played a crucial role in the development of AlphaFold, particularly in refining the **deep neural network** architecture that underpins the system's ability to predict protein structures. Jumper's background in **biophysics** and **computational biology** enabled him to bridge the gap between AI and structural biology, ensuring that AlphaFold's predictions were not only accurate but also biologically meaningful.

One of the key innovations that Jumper introduced to AlphaFold was the use of **attention mechanisms**, a technique that allows the AI to focus on the most relevant parts of a protein sequence when making predictions. This approach enabled AlphaFold to model the complex interactions between distant regions of a protein, which are critical for determining its final structure. Jumper's contributions to the **CASP14** competition, where AlphaFold outperformed all other methods, solidified his reputation as one of the leading figures in the field of **protein folding**.

Jumper's collaboration with Hassabis and Baker highlights the importance of interdisciplinary teamwork in advancing scientific discovery. By bringing together expertise in **machine learning**, **biophysics**, and **protein design**, they were able to achieve results that would have been impossible for any one individual or team working in isolation.

Collaborative Efforts in Protein Structure Prediction

The collaboration between Hassabis, Baker, and Jumper represents a **synergy** between **AI-driven** and **physics-based** approaches to protein structure prediction. Each of these researchers brought unique skills and perspectives to the table, enabling them to solve problems that had stymied the scientific community for decades.

1. **AI-Driven Insights (Hassabis and Jumper)**: AlphaFold's deep learning approach allows it to learn patterns from large datasets of known protein structures, making it highly effective at predicting the structures of new proteins. This method has proven particularly valuable in cases where traditional experimental methods are difficult or impossible to apply.

2. **Physics-Based Simulations (Baker)**: Rosetta's physics-based approach complements AlphaFold by providing insights into the molecular interactions that drive protein folding. This allows researchers to design proteins with specific properties, such as increased stability or enhanced binding affinity for a particular target.

3. **Real-World Applications**: The collaboration between these researchers has led to real-world applications in **medicine**, **biotechnology**, and **environmental science**. For example, their work has contributed to the development of **therapeutics** for diseases such as **COVID-19**, as well as the design of **enzymes** that can break down **plastic waste** or produce **biofuels** more efficiently.

Shaping Future Biotechnological Innovations

The collaboration between Hassabis, Jumper, and Baker is not just a success story of past achievements; it is a **blueprint** for future innovations in biotechnology. By combining the strengths of AI-driven and physics-based approaches, these researchers are paving the way for new discoveries in fields such as **personalized medicine**, **synthetic biology**, and **environmental sustainability**.

1. **Personalized Medicine**: One of the most exciting applications of protein structure prediction is in the field of **personalized medicine**. By predicting the structures of proteins involved in genetic diseases, researchers can design **targeted therapies** that are tailored to an individual's unique genetic makeup. This approach has the potential to revolutionize the

treatment of diseases such as cancer, where **mutations** in specific proteins drive tumor growth.

2. **Synthetic Biology**: In the field of **synthetic biology**, AlphaFold and Rosetta are being used to design entirely new proteins with specific functions. This includes enzymes that can catalyze chemical reactions, proteins that can bind to specific molecules, and even proteins that can perform tasks not found in nature. These innovations have applications in **biotechnology**, **medicine**, and **agriculture**, offering new solutions to global challenges such as **climate change** and **food security**.

3. **Environmental Sustainability**: The ability to design proteins with specific functions also has important implications for **environmental sustainability**. For example, researchers are using AlphaFold and Rosetta to design enzymes that can break down **plastic waste**, reducing the environmental impact of plastic pollution. Similarly, these systems are being used to develop **biofuels** that are more efficient and sustainable than traditional fossil fuels.

The Nobel Prize and the Future of Collaboration in Science

The awarding of the **2024 Nobel Prize in Chemistry** to **Demis Hassabis**, **John Jumper**, and **David Baker** is a testament to the power of **collaboration** in scientific research. By combining their expertise in **artificial intelligence** and **biochemistry**, these researchers were able to solve one of the most complex problems in molecular biology and open new avenues for scientific discovery.

Their work has demonstrated that **interdisciplinary collaboration** is essential for addressing the challenges of the 21st century. The problems we face—whether in **medicine**, **environmental science**, or **biotechnology**—are too complex for any one discipline to solve on its own. By bringing together experts from diverse fields, we can achieve breakthroughs that would have been unimaginable just a few years ago.

Looking ahead, the collaboration between Hassabis, Jumper, and Baker is likely to inspire further innovations in biotechnology. As AI systems like AlphaFold continue to evolve, they will become evenfor solving the world's most pressing challenges. As Hassabis, Baker, and Jumper continue their collaborative work, the scientific community can look forward to further breakthroughs in **AI-driven biological research**, with implications that extend far beyond protein structure prediction.

The Nobel Prize, awarded to these researchers, underscores the **growing recognition** of **artificial intelligence** as a critical tool in scientific discovery. It signals a shift in how the scientific community views AI—not just as a technological innovation but as an essential component of modern research, capable of solving problems that were once considered intractable.

The collaboration between **Demis Hassabis**, **David Baker**, and **John Jumper** represents a new frontier in **scientific innovation**, driven by the intersection of **artificial intelligence** and **biochemistry**. By combining the **deep learning** capabilities of AlphaFold with the **physics-based simulations** of Rosetta, these researchers have revolutionized the field of **protein structure prediction**. Their work has not only earned them the **2024 Nobel Prize in Chemistry** but also opened new possibilities for **personalized medicine**, **synthetic biology**, and **environmental sustainability**.

As we look to the future, the partnership between AI and traditional scientific methods will continue to drive advancements in **biotechnology** and other fields. The success of this collaboration serves as a model for how interdisciplinary teams can achieve extraordinary results, solving some of the most complex challenges facing humanity today. The innovations brought about by Hassabis, Baker, and Jumper will undoubtedly have a lasting impact on science, medicine, and technology, shaping the future of research and discovery for years to come.

CHAPTER 9
The Role of Artificial Intelligence in Future Scientific Discoveries

The **role of artificial intelligence (AI)** in advancing scientific research has already been firmly established through groundbreaking innovations like **AlphaFold**, which has revolutionized **protein structure prediction**. The success of AI in biological research signals a new era where machines not only augment human ingenuity but also drive progress in fields previously hindered by technological or methodological limitations. As we look to the future, the potential for AI to transform other scientific domains—such as **chemistry**, **physics**, **materials science**, and even **climate research**—seems boundless.

This chapter will explore the future of AI in scientific research, speculate on how AI could solve other complex scientific challenges, and examine how leaders like **Demis Hassabis** envision AI's expanding role as a tool for **interdisciplinary research** and **global problem-solving**.

AI in Scientific Discovery: A New Frontier

The application of AI in scientific research is still in its early stages, but it has already demonstrated its capacity to accelerate discoveries in ways that were previously unimaginable. Traditional scientific methods rely heavily on **hypothesis-driven research**, which is limited by the pace of human experimentation and the sheer complexity of many scientific problems. **AI-driven research**, on the other hand, can sift through vast amounts of data, identify patterns, and generate new hypotheses much faster than human researchers.

In **biological research**, AlphaFold's ability to predict protein structures from amino acid sequences provides a clear example of AI's transformative power. For over 50 years, the **protein folding problem** had stymied scientists, limiting progress in **drug discovery**, **genomics**, and **molecular biology**. AlphaFold's success in solving this problem showcases AI's ability to not only enhance traditional methods but also to generate insights that humans might struggle to uncover due to the complexity and scale of the data involved.

The question now is: What other scientific challenges might AI help to solve in the future?

AI's Role in Biology: Beyond Protein Structure Prediction

Beyond the successes of AlphaFold, AI's role in **biology** is expected to expand into a wide range of areas. One of the most promising applications lies in **genomics** and **personalized medicine**. AI is already being used to analyze **genomic data** to identify patterns that could predict disease risk, understand the genetic basis of complex diseases like **cancer**, and guide the development of targeted therapies. By leveraging AI's ability to process and analyze large datasets, researchers hope to tailor treatments to individual patients, offering **personalized therapies** that are more effective and have fewer side effects than traditional "one-size-fits-all" treatments.

Another area where AI is poised to have a major impact is in **synthetic biology**. Scientists are increasingly using AI to design new biological systems, such as **custom enzymes** or **genetically modified organisms** that can perform specific tasks, like breaking down pollutants or producing biofuels. AI systems can rapidly simulate the behavior of these synthetic systems, allowing researchers to test thousands of designs in silico before choosing the most promising candidates for real-world experiments. This approach could revolutionize industries ranging from **biotechnology** to **agriculture**, enabling the development of new crops, drugs, and sustainable materials.

Finally, AI has the potential to accelerate the study of **neuroscience** and **brain function**. Understanding the human brain—arguably the most complex structure in the universe—remains one of the great frontiers of science. AI could help researchers model the intricate networks of neurons and synapses that underlie thought, memory, and consciousness. By simulating the brain's structure and function, AI could lead to breakthroughs in understanding **neurodegenerative diseases** like Alzheimer's or in developing **brain-computer interfaces (BCIs)** that allow humans to control machines with their thoughts.

AI in Chemistry: Discovering New Materials

The field of **chemistry** has long been associated with painstaking experiments, where researchers synthesize new compounds and test their properties one by one. AI, however, is poised to revolutionize **materials discovery** by enabling researchers to predict the properties of new compounds before they are synthesized in the lab.

In recent years, AI has been used to model **chemical reactions**, predict the properties of **molecules** and

materials, and even design entirely new compounds. This approach is particularly valuable in fields like **drug discovery**, where researchers must evaluate thousands or even millions of potential molecules to find those that are effective against specific diseases. AI-driven **virtual screening** allows scientists to test these molecules in silico, saving time and resources by focusing experimental efforts on the most promising candidates.

One of the most exciting applications of AI in chemistry is the discovery of **new materials** with properties that can be tailored to specific applications. For example, AI is being used to design **battery materials** that can store more energy, **catalysts** that speed up industrial reactions, and **superconductors** that can transmit electricity with zero resistance. These materials have the potential to revolutionize industries ranging from **renewable energy** to **electronics**.

AI is also being applied to solve long-standing challenges in **quantum chemistry**, where researchers are working to model the behavior of atoms and molecules at the quantum level. AI-driven models are helping scientists understand the properties of complex chemical systems that were previously too difficult to study, leading to the discovery of new **drugs**, **materials**, and **chemical processes**.

AI in Physics: Understanding the Universe

While AI has already made significant contributions to biology and chemistry, its potential impact on **physics** is equally promising. In particular, AI could help researchers tackle some of the most challenging problems in **theoretical physics**, such as understanding the nature of **dark matter**, **dark energy**, and the fundamental laws governing the universe.

In recent years, AI has been used to analyze data from **particle accelerators**, such as the **Large Hadron Collider (LHC)**, to search for new **subatomic particles** and **fundamental forces**. AI systems are particularly well-suited for this type of research because they can process vast amounts of data and identify patterns that are difficult for human researchers to detect. By automating the process of data analysis, AI has the potential to accelerate discoveries in **high-energy physics**, potentially leading to new insights into the building blocks of matter and energy.

AI could also play a key role in understanding **cosmology** and the structure of the universe. For example, AI is being used to analyze data from **telescope surveys** to map the distribution of galaxies and understand how they evolved over billions of years. By analyzing these data, AI can help researchers test theories about the nature of **dark matter** and **dark energy**, which together make up more than 95% of the universe's mass-energy content but remain poorly understood.

In addition to its role in theoretical physics, AI is also being used to model **physical systems** in areas such as **fluid dynamics**, **quantum mechanics**, and **astrophysics**. These models can simulate the behavior of complex systems that are difficult or impossible to study experimentally, leading to new insights into phenomena ranging from **black holes** to **quantum entanglement**.

AI and Climate Change: A Tool for Environmental Solutions

One of the most pressing challenges facing humanity today is **climate change**, and AI is emerging as a powerful tool for addressing this global crisis. By analyzing vast amounts of data from **climate models**, **satellite observations**, and **weather sensors**, AI systems can help researchers understand how the climate is changing and predict the future impacts of global warming.

AI is being used to optimize **energy systems**, such as **smart grids** and **renewable energy infrastructure**, making them more efficient and sustainable. For example, AI algorithms can predict fluctuations in energy demand and adjust power generation in real-time, ensuring that renewable energy sources like **solar** and **wind** are used to their full potential. In addition, AI can be used to design new materials for **energy storage**, such as **batteries** that can store more energy for longer periods, making renewable energy more reliable.

AI is also playing a key role in **environmental monitoring** and **conservation** efforts. By analyzing data from **satellite images**, AI systems can monitor **deforestation**, **wildlife populations**, and **air quality**, helping governments and organizations track the health of ecosystems and respond to environmental threats more quickly. For example, AI is being used to predict the spread of **wildfires** and optimize **forest management** practices to reduce the risk of catastrophic fires.

In agriculture, AI is helping farmers adapt to the challenges posed by **climate change** by optimizing **irrigation**, **fertilizer use**, and **crop selection**. By analyzing weather data and soil conditions, AI systems can recommend the best practices for increasing crop yields while reducing water use and minimizing environmental impact. These AI-driven solutions are essential for ensuring food security in a changing climate.

Demis Hassabis' Vision for AI in Interdisciplinary Research

Demis Hassabis has long been an advocate for the role of AI in **interdisciplinary research**, believing that AI has the potential to solve some of the world's most complex problems by bridging the gaps between **different scientific disciplines**. Hassabis has emphasized that the future of AI lies not in replacing human researchers, but in **augmenting human creativity** and helping scientists tackle problems that were previously considered unsolvable.

One of Hassabis' key beliefs is that AI can help researchers work more **collaboratively**, bringing together experts from fields as diverse as **neuroscience**, **molecular biology**, **materials science**, and **climate research**. By enabling scientists to share data and insights more easily, AI can facilitate interdisciplinary collaboration and accelerate the pace of discovery.

In addition, Hassabis and his team have demonstrated that AI can serve as a **unifying force** in science, allowing researchers to approach problems from multiple perspectives simultaneously. By integrating knowledge from diverse fields, AI can uncover new connections and generate insights that were previously hidden.

For example, AI-driven research in **neuroscience** could help biologists understand the mechanisms of **memory** and **learning**, which could, in turn, inform the design of **artificial neural networks** for **machine learning**. Similarly, AI systems designed for **materials discovery** could provide insights that are applicable to **quantum chemistry** or **environmental science**, leading to breakthroughs in **energy storage** or **carbon capture** technologies.

As AI continues to evolve, Hassabis envisions a future where AI acts as a **catalyst for innovation** across all scientific disciplines. By enabling researchers to process larger datasets, test more hypotheses, and simulate more complex systems than ever before, AI has the potential to transform not only individual fields of study but also the entire scientific enterprise.

The future of **artificial intelligence** in scientific research holds immense promise, with applications that extend far beyond the breakthroughs already achieved in **biology**, **chemistry**, and **physics**. As AI systems become more sophisticated, they will continue to accelerate the pace of discovery in fields ranging from **medicine** to **climate science**, offering new solutions to some of the world's most pressing challenges.

Demis Hassabis and his collaborators have demonstrated how AI can be harnessed to solve complex problems, such as the **protein folding problem**, through interdisciplinary collaboration. As AI continues to develop, it will become an even more essential tool for **global problem-solving**, helping researchers tackle challenges like **climate change**, **disease**, and **sustainability**.

Looking ahead, the role of AI in science will not be limited to a single discipline or application. Instead, AI will serve as a **bridge** between different fields of research, enabling scientists to work together in new and innovative ways. As AI continues to evolve, its potential to **augment human intelligence** and transform the scientific process will be limited only by our imagination.

CHAPTER 10
Demis Hassabis' Legacy and Future Prospects

Demis Hassabis stands as one of the most influential figures in the field of **artificial intelligence (AI)**, having spearheaded innovations that not only transformed the technology sector but also significantly impacted science, medicine, and global problem-solving. His work, particularly with **DeepMind**, has established a new frontier in AI, and his contributions have been instrumental in addressing challenges that once seemed insurmountable. From the revolutionary developments of **AlphaGo** and **AlphaFold** to his visionary approach to AI ethics and interdisciplinary research, Hassabis has consistently pushed the boundaries of what AI can achieve.

In this chapter, we will reflect on Hassabis' legacy, examining the profound impact his work has had on both **science** and **technology**. Additionally, we will explore the future prospects of his career, speculating on the contributions he may continue to make in **AI, biology**, and beyond. Finally, we will consider how his work could inspire the next generation of scientists and AI researchers to expand the horizons of human knowledge and technology.

Demis Hassabis: A Pioneering Figure in Artificial Intelligence

Demis Hassabis' legacy is defined by his role as a **pioneer** in the development and application of AI to solve some of the most complex problems in science and technology. Born in **London** to a mixed heritage family, Hassabis' intellectual journey began at a young age. A **child prodigy** in chess, he became a master by the age of 13 and later studied **computer science** at the **University of Cambridge**. His early interests in **video games**, **chess**, and **neuroscience** foreshadowed the interdisciplinary approach he would later bring to his AI research.

Hassabis co-founded **DeepMind** in **2010** with the vision of building **general artificial intelligence**—a machine that could mimic the learning and problem-solving abilities of the human brain. His background in **neuroscience**, combined with his expertise in **AI**, gave him a unique perspective on how to approach the development of intelligent systems. He believed that understanding the workings of the brain could help scientists design AI systems capable of learning and generalizing across domains.

The early success of DeepMind's **game-playing AIs**, particularly **AlphaGo**, propelled Hassabis into the global spotlight. AlphaGo, an AI system designed to play the ancient board game **Go**, achieved a historic milestone in 2016 when it defeated **Lee Sedol**, one of the world's top Go players. This victory was significant because Go, with its vast number of possible moves and complex strategy, had long been considered beyond the reach of computers. AlphaGo's success demonstrated that **deep learning** and **reinforcement learning** techniques could surpass human expertise in highly complex domains.

However, Hassabis' ambitions extended far beyond games. He envisioned AI as a tool that could solve real-world problems, and this vision became a reality with the development of **AlphaFold**, an AI system capable of predicting **protein structures**. The **protein folding problem**, which had perplexed scientists for decades, was solved with AlphaFold's breakthrough in 2020, when the system achieved **near-experimental accuracy** in predicting protein structures based on their amino acid sequences. This achievement has had a profound impact on **biomedicine**, **drug discovery**, and **genomics**, opening new avenues for scientific research and therapeutic development.

Hassabis' legacy is thus built on his ability to apply **AI** to diverse and critical areas of science, transforming fields such as **neuroscience, molecular biology,** and **genomics**. His work has demonstrated that AI can be a powerful tool for **solving problems** that require understanding vast amounts of complex data, from predicting protein structures to discovering new **therapeutics** for diseases.

Impact on Science and Technology

Hassabis' contributions to **AI** and its applications in science have had far-reaching consequences, influencing the way researchers approach some of the most pressing challenges in **biological science, medicine,** and **technology**.

1. **AlphaFold and Protein Structure Prediction**: Perhaps Hassabis' most significant scientific contribution is the development of **AlphaFold**, a system that uses AI to predict the **three-dimensional structures** of proteins. This has had a profound impact on **biology** and **medicine**, as understanding protein structures is critical for developing new **drugs, vaccines,** and **therapies** for diseases ranging from **cancer** to **Alzheimer's disease**. By making AlphaFold's protein structure predictions freely available to the global scientific community,

Hassabis has enabled researchers worldwide to accelerate their research and develop new treatments for previously intractable conditions.

2. **AI in Healthcare and Drug Discovery**: Beyond AlphaFold, Hassabis has driven efforts to apply **AI** to other areas of **biomedicine**. DeepMind's AI systems are being used to analyze **medical images**, predict patient outcomes, and assist doctors in diagnosing diseases. For example, DeepMind's collaboration with **Moorfields Eye Hospital** in **London** led to the development of an AI system capable of diagnosing **eye diseases** with the same level of accuracy as human doctors. AI is also being applied to **drug discovery**, where it can rapidly screen **molecular compounds** and predict their interactions with biological targets, significantly speeding up the development of new drugs.

3. **Ethical AI Development**: Hassabis has been an outspoken advocate for the **ethical development** of AI. He has emphasized the importance of ensuring that AI is used responsibly and benefits all of humanity. Under his leadership, DeepMind established an **AI Ethics Board** to oversee the

ethical implications of its research and ensure that its technologies are used for socially beneficial purposes. Hassabis has also been involved in initiatives to **regulate AI development**, particularly in the context of **autonomous weapons** and the potential misuse of AI for **surveillance**.

4. **Interdisciplinary Research**: One of Hassabis' most important contributions to science is his commitment to **interdisciplinary research**. By bringing together experts from fields as diverse as **neuroscience**, **machine learning**, and **biochemistry**, he has demonstrated that the most significant scientific advances often occur at the intersection of different disciplines. His approach to **AI research** has inspired new collaborations between computer scientists, biologists, and medical researchers, fostering innovation across a range of scientific fields.

Hassabis' Future Contributions to AI and Biology

Looking ahead, **Demis Hassabis** is poised to continue making significant contributions to both **AI** and **biology**. His work with **DeepMind** is far from over, and there are several areas where Hassabis' leadership and vision could drive future innovations.

1. **Advancing General AI**: One of Hassabis' long-term goals has been to develop **general artificial intelligence (AGI)**—a machine that can perform any intellectual task that a human can do. While current AI systems, such as **AlphaFold** and **AlphaGo**, are highly specialized, the development of AGI would represent a major leap forward in AI research. Hassabis has suggested that **understanding the human brain** will be key to developing AGI, and his background in **neuroscience** positions him uniquely to tackle this challenge. In the future, we may see breakthroughs in AGI that could revolutionize **robotics**, **education**, **healthcare**, and many other sectors.

2. **AI in Personalized Medicine**: Hassabis' work on **AlphaFold** and other AI-driven systems has already demonstrated the potential for **personalized medicine**, where treatments are tailored to an individual's genetic makeup. In the future, AI could be used to analyze a patient's **genomic data**, predict their risk for certain diseases, and develop personalized therapies based on their unique genetic profile. This approach has the potential to

revolutionize healthcare, providing more effective treatments with fewer side effects.

3. **AI for Environmental Sustainability**: As the world grapples with the challenges of **climate change** and **environmental degradation**, AI is likely to play a key role in developing solutions. Hassabis has expressed interest in using AI to address global challenges such as **sustainable energy**, **biodiversity conservation**, and **resource management**. AI systems could be used to model **ecosystems**, predict the effects of **climate change**, and optimize the use of natural resources, helping governments and organizations make more informed decisions about how to protect the environment.

4. **Expanding AI's Role in Scientific Discovery**: As AI continues to evolve, Hassabis is likely to play a leading role in expanding its applications in scientific research. From **quantum physics** to **materials science**, AI has the potential to accelerate discoveries in fields that were once considered too complex for traditional methods. By partnering with researchers across disciplines, Hassabis could help unlock new

scientific breakthroughs that could transform industries ranging from **energy** to **biotechnology**.

Inspiring the Next Generation of AI Researchers

Demis Hassabis' legacy extends beyond his individual achievements—he is also an inspiration to the next generation of **AI researchers** and **scientists**. Through his leadership at **DeepMind**, Hassabis has demonstrated that **curiosity**, **interdisciplinary thinking**, and a commitment to **ethical AI development** are essential qualities for those seeking to make meaningful contributions to science and technology.

1. **Mentorship and Collaboration**: One of the most significant ways Hassabis is inspiring future researchers is through his **mentorship** and **collaborative approach** to research. At DeepMind, he has fostered a culture of collaboration, where experts from different fields work together to solve complex problems. By emphasizing the importance of **interdisciplinary research**, Hassabis has shown that the next collaboration and cross-disciplinary work, he is preparing the next generation of scientists to tackle complex challenges that require diverse expertise.

2. **Commitment to AI Ethics**: Hassabis' emphasis on the ethical development of AI is likely to shape the future of AI research for years to come. By promoting transparency, fairness, and accountability in AI development, he has set a standard for how AI should be used to benefit society. As AI becomes more pervasive in everyday life, future researchers will need to grapple with the ethical implications of their work, and Hassabis' approach provides a valuable framework for doing so.

3. **Public Engagement and Education**: Through his public talks, interviews, and writings, Hassabis has played a key role in **educating the public** about AI and its potential. By explaining complex concepts in accessible ways, he has helped demystify AI and inspired young people to pursue careers in science and technology. His vision of AI as a tool for **global problem-solving** has captured the imagination of many, and his efforts to promote **ethical AI** have helped build public trust in the technology.

Conclusion: A Legacy of Innovation and Inspiration

Demis Hassabis' legacy as a **pioneer** in AI is already well-established, but his work is far from finished. His contributions to **science** and **technology** have already had a profound impact on fields ranging from **biology** to **medicine**, and his commitment to **ethical AI development** has set a standard for how AI should be used responsibly. As AI continues to evolve, Hassabis is likely to remain at the forefront of innovation, driving new breakthroughs in **general AI**, **personalized medicine**, and **environmental sustainability**.

Beyond his own achievements, Hassabis' greatest legacy may be the **inspiration** he provides to the next generation of **AI researchers** and **scientists**. By promoting **collaboration**, **interdisciplinary research**, and **ethical responsibility**, he has shown that the greatest scientific advances come from **curiosity**, **creativity**, and a commitment to improving the world. As we look to the future, it is clear that **Demis Hassabis** will continue to play a central role in shaping the direction of AI and its applications, inspiring others to push the boundaries of what's possible.

CONCLUSION

Demis Hassabis and the Nobel Prize – A Legacy in the Making

As the story of **Demis Hassabis** and his groundbreaking work unfolds, it becomes clear that he stands among the most influential pioneers of **artificial intelligence (AI)** and its application to solving some of the world's most complex scientific problems. From his childhood as a **chess prodigy** to his time as a **neuroscience student** at **Cambridge University** and later as a **co-founder of DeepMind**, Hassabis has consistently shown an ability to bridge fields, leveraging the power of **computational intelligence** to redefine science and technology.

This book has chronicled Hassabis' journey, from the creation of **AlphaGo**, which shocked the world by defeating a top-ranked Go player, to **AlphaFold**, the AI that solved the protein-folding problem, a challenge that had confounded scientists for half a century. It is these significant contributions that have solidified Hassabis' legacy as a visionary in **machine learning** and **deep learning**, reshaping how we view both **artificial intelligence** and its applications in **biological science**.

Hassabis' potential to win a **Nobel Prize**—whether in **Chemistry**, for his role in revolutionizing **protein structure prediction**, or in another category yet to be created for **artificial intelligence**—seems not only possible but inevitable. The world is witnessing a turning point in how AI systems are not only supporting but actively leading the charge in scientific discovery. This chapter serves as a reflection on the key themes of Hassabis' career, the global impact of his work, and what the future holds for him and the broader scientific community.

The Transformative Impact of Artificial Intelligence on Science

The **Nobel Prize** represents the pinnacle of recognition for scientific achievement. Historically, the prize has been awarded for groundbreaking discoveries in fields such as **physics, chemistry, medicine**, and **literature**. While AI has yet to receive a Nobel, its profound impact on scientific research is undeniable, and **Demis Hassabis** is a significant driver of that impact.

At the core of Hassabis' legacy is the understanding that **AI is not just a tool**; it is becoming an essential partner in the process of scientific discovery. In the past, AI was largely seen as a method to automate routine tasks or process large datasets. However, under Hassabis' leadership, DeepMind has demonstrated that AI can push the boundaries of what we know and even **solve previously unsolvable problems**. This redefines AI's role in science, shifting from assistance to **active discovery**.

Take, for instance, **AlphaFold**, a system that effectively solved the **protein folding problem** by predicting the 3D structure of proteins from their amino acid sequences with near-experimental accuracy. Protein folding is central to understanding biological processes and developing drugs, and AlphaFold has opened new pathways for research in **medicine**, **genomics**, and **biochemistry**. This accomplishment alone has far-reaching implications, as it accelerates drug discovery, offers insights into **disease mechanisms**, and facilitates the design of new **biomaterials**. In these ways, Hassabis' work is likely to improve the lives of countless people by helping to address global health challenges.

Hassabis' legacy will also be measured by his vision for AI's role in **interdisciplinary research**. He has successfully combined his knowledge of **neuroscience**, **game theory**, and **machine learning** to create systems that reflect not only computational efficiency but also biological intelligence. By showing how AI can simulate aspects of human cognition, Hassabis has encouraged a multidisciplinary approach to AI development, one that involves experts from various fields working together to solve **real-world problems**. This will no doubt influence future generations of AI researchers and scientists.

Demis Hassabis' Nobel Prize Worthy Contributions

While **Demis Hassabis** has not yet won a Nobel Prize, his achievements position him as a strong contender in the near future. The **Nobel Committee** has a long history of awarding prizes to scientists whose work has had profound, far-reaching impacts, and the breakthrough represented by AlphaFold aligns with this tradition. The system's ability to solve one of biology's oldest problems, the accurate prediction of protein structures, was hailed as a "game-changer" by many in the scientific community, including Nobel laureates. It is not far-fetched to imagine that the committee could recognize AI's role in transforming **molecular biology** and award Hassabis a **Nobel Prize in Chemistry**.

Beyond AlphaFold, Hassabis' work on **AlphaGo**, **deep reinforcement learning**, and the development of **general AI** has reshaped how industries view the potential of AI systems. His innovations have implications for many fields, from **autonomous systems** and **healthcare** to **climate science** and **education**. As AI continues to play a more significant role in **scientific discovery**, it is possible that the **Nobel Prize categories** themselves may expand to include **artificial intelligence**, recognizing its revolutionary role in **scientific methodology**.

Hassabis' contributions are particularly notable in the context of **biology** and **chemistry**. By accelerating the discovery of protein structures, AlphaFold has the potential to vastly improve the process of **drug design**, especially for diseases that have been resistant to traditional approaches. AlphaFold's data is being used to understand proteins related to **Alzheimer's**, **Parkinson's**, **cancer**, and **COVID-19**. Should AI-driven advances continue to yield real-world medical breakthroughs, the Nobel Prize would be a fitting acknowledgment of the foundational role that Hassabis and his team have played.

The Future of AI and Science: Hassabis' Continuing Legacy

As we look to the future, **Demis Hassabis** and **DeepMind** remain at the forefront of AI innovation. Hassabis has stated that his goal is to develop **general artificial intelligence**

(**AGI**)—a system that can perform any intellectual task a human being can do. While this goal remains a work in progress, the strides made by systems like **AlphaFold** show that we are closer than ever to achieving more versatile, adaptive AI systems capable of making independent, novel discoveries.

In the coming years, AI is likely to have an increasing impact on a variety of scientific fields. AI could be used to model **quantum systems**, predict the effects of **climate change**, optimize **energy systems**, and design entirely new materials through **computational chemistry**. Given his track record, Hassabis will likely play a leading role in these developments, positioning AI as an indispensable tool in **problem-solving** across disciplines.

Moreover, Hassabis' work has paved the way for **ethical AI development**, a subject that will grow increasingly important as AI becomes more integrated into our daily lives. His emphasis on **transparency, accountability**, and the **social responsibility** of AI researchers has set a precedent for how AI should be developed and used. As AI systems begin to take on more roles in **healthcare, finance, law**, and **national security**, the ethical frameworks advocated by Hassabis will become even more critical.

Hassabis' continuing influence will extend to **mentorship** as well. Through DeepMind and collaborations with **universities** and **research institutes**, Hassabis has mentored the next generation of **AI researchers**, shaping the future of the field. As more young scientists enter the field of AI, inspired by Hassabis' accomplishments, his legacy will be further extended, ensuring that his innovations have a lasting impact on **scientific progress** and **technological development**.

Inspiring the Next Generation

Perhaps one of Hassabis' greatest legacies will be the inspiration he provides to future scientists and AI researchers. His work demonstrates that **curiosity**, **creativity**, and an openness to interdisciplinary collaboration can lead to transformative breakthroughs. By pushing the boundaries of **machine learning** and **deep learning**, Hassabis has shown that **artificial intelligence** is not limited to theoretical applications; it can actively solve real-world problems that have stumped generations of scientists.

Young scientists, inspired by Hassabis, may take up the challenge of **combining AI with biology, physics, medicine**, and even **social sciences** to solve global problems such as **climate change**, **pandemics**, and **energy shortages**. Hassabis has shown that the future of science lies in collaboration between fields, and that the next great breakthroughs will likely come from teams of experts working across traditional disciplinary boundaries.

Conclusion: A Visionary in Science and Technology

Demis Hassabis has already cemented his place as a **visionary** in **artificial intelligence**. His work has redefined what is possible in the fields of **biology, medicine**, and **technology**, bringing AI from the realm of theory into real-world applications that are transforming science. The development of **AlphaGo**, **AlphaFold**, and DeepMind's broader initiatives have shown the world that AI can not only match but exceed human capabilities in certain complex tasks, opening new avenues for discovery and innovation.

As Hassabis continues his work, it is clear that his legacy will only grow. Whether or not he wins a **Nobel Prize**, his contributions to **AI, neuroscience**, and **interdisciplinary research** will have a lasting impact on the scientific community and the world at large. He has paved the way for future generations of scientists to continue pushing the boundaries of what AI can achieve, and his ethical approach to AI development ensures that these technologies will be used for the benefit of all.

www.ingramcontent.com/pod-product-compliance
Lightning Source LLC
Chambersburg PA
CBHW050305230526
45471CB00005B/2034